# PHYSICS OF METALS
## 2. DEFECTS

# THE
# PHYSICS OF METALS

## 2. DEFECTS

---

EDITED BY
## P. B. HIRSCH

CAMBRIDGE UNIVERSITY PRESS

CAMBRIDGE

LONDON · NEW YORK · MELBOURNE

CAMBRIDGE UNIVERSITY PRESS
Cambridge, New York, Melbourne, Madrid, Cape Town, Singapore, São Paulo, Delhi

Cambridge University Press
The Edinburgh Building, Cambridge CB2 8RU, UK

Published in the United States of America by Cambridge University Press, New York

www.cambridge.org
Information on this title: www.cambridge.org/9780521113106

First published 1975
This digitally printed version 2009

A catalogue record for this publication is available from the British Library

Library of Congress Catalogue Card Number: 74–14439

ISBN 978-0-521-20077-6 hardback
ISBN 978-0-521-11310-6 paperback

# PREFACE

By convention a *Festschrift* is a collection of pretty 'essays' by a miscellany of scholars, presented to an older and greater scholar at some significant milestone of his life. The present work conforms to this convention only in that it was conceived as a tribute to Professor Sir Nevill Mott upon the occasion of his 60th birthday.

The prime intention was to write a modern version of 'Mott and Jones' – a comprehensive and up-to-date account of the Theory of Metals and Alloys. But much has been discovered in 30 years, and this would have stretched to many thousands of pages. We therefore decided to concentrate on the two major topics where our interests mainly lie – the electronic structure and electrical properties of metals, on the one hand, and the mechanical properties of solids on the other. In the end, each topic grew and diverged into a separate book.

The first volume, *Electrons*, published in 1969, gave an account of current understanding of the electron theory of metals, with particular reference to band structure, Fermi surfaces and transport properties. The present, second, volume deals with lattice defects and the mechanical properties of metals and alloys. Although the titles of the various chapters (Point defects, Crystal dislocations, Observations of defects in metals by electron microscopy, Solution and precipitation hardening, Work hardening, and Fracture) suggest a systematic coverage of a very large field in Physics and Metallurgy, the book is certainly not intended to be a comprehensive review of the subject. Instead, the individual contributors have generally concentrated on certain aspects of their chosen area, giving a coherent and detailed account of a limited range of topics. The book is aimed at research workers and advanced students. The emphasis has been on the physics of the phenomena, rather than on detailed mathematical theory, as in much of Mott's own work. Mott's influence is apparent not only in the approach and work of the various authors, six of whom have had the privilege of working with and being stimulated by him at Bristol or at Cambridge, but of course also in the many contributions which he has made in this wide field, in which he has always aimed at explaining macroscopic phenomena in terms of atomic or defect mechanisms.

It is unfortunate that the publication of this second volume has been so long delayed. Some of the contributions were prepared as long ago as 1966; the last one was received in 1971. The editor must take full responsibility for this delay. In some cases some limited updating of the contributions has been carried out. In spite of the absence of more recent references, however, the work should be of value partly because of the definitive nature of major sections of the book, and partly because of the individual approach of the contributors, many of whom have pioneered and contributed in an original way to the subject.

Despite their labours, the authors of this volume, like those of volume 1, will not receive any royalties, which will all be paid over into a trust fund to encourage study and expertise in foreign languages amongst professional scientists. This arrangement, which was suggested by Mott himself, corresponds most felicitously with his own interest in languages and with the wide international circle of his friendships.

It is our duty to record grateful acknowledgment to authors and publishers who have given permission for the reproduction of various diagrams and micrographs in this book. The editor would also like to thank the other contributors for accepting so graciously the long delay in publication of their work.

P. B. H.

# CONTENTS

# SIR NEVILL MOTT

MOTT, *Sir Nevill (Francis)* Kt 1962; FRS 1936; MA Cantab; Cavendish Professor of Experimental Physics, 1954–71; Senior Research Fellow, Imperial College, London, since 1971; b. 30 Sept. 1905; s. of C.F. Mott, late Dir of Educn, Liverpool, and Lilian Mary Reynolds; m. 1930, Ruth Horder; two d. *Educ*: Clifton Coll.; St John's Coll., Cambridge. Lecturer at Manchester Univ., 1929–30; Fellow and Lecturer, Gonville and Caius Coll., Cambridge, 1930–33; Melville Wills Prof. of Theoretical Physics in the Univ. of Bristol, 1933–48; Henry Overton Wills Prof. and Dir of the Henry Herbert Wills Physical Laboratories, Univ. of Bristol, 1948–54. Master of Gonville and Caius Coll., Univ. of Cambridge, 1959–66. Hughes Medal of Royal Society, 1941, Royal Medal, 1953; corr. mem., Amer. Acad. of Arts and Sciences, 1954; Pres., International Union of Physics, 1951–57; Pres., Mod. Languages Assoc., 1955; Pres., Physical Soc., 1956–58; Mem. Governing Board of Nat. Inst. for Research in Nuclear Science, 1957–60; Mem. Central Advisory Council for Education for England, 1956–59; Mem. Academic Planning Cttee and Council of University Coll. of Sussex; Chm. Ministry of Education's Standing Cttee on Supply of Teachers, 1959–62; Mem., Inst. of Stragetic Studies; Chm. Nuffield Foundation's Cttee on Physics Education, 1961; Physics Education Cttee (Royal Society and Inst. of Physics), 1965. Foreign Associate, Nat. Acad. of Sciences of USA, 1957; Hon. Member: Akademie der Naturforscher Leopoldina, 1964; Société Française de Physique, 1970. Hon. DSc (Louvain, Grenoble, Paris, Poitiers, Bristol, Ottawa, Liverpool, Reading, Sheffield, London, Warwick, Lancaster). Grande médaille de la Société Française de Métallurgie, 1970. Publications: *An Outline of Wave Mechanics*, 1930; *The Theory of Atomic Collisions* (with H. S. W. Massey), 1933; *The Theory of the Properties of Metals and Alloys* (with H. Jones), 1936; *Electronic Processes in Ionic Crystals* (with R. W. Gurney), 1940; *Wave Mechanics and its Applications* (with I. N. Sneddon), 1948; *Elements of Wave Mechanics*, 1952; *Atomic Structure and the Strength of Metals*, 1956; *Electronic Processes in Non-Crystalline Materials* (with E. A. Davis), 1971; various contribs to scientific periodicals about Atomic Physics, Metals, Semiconductors and Photographic Emulsions. Address: The Cavendish Laboratory, Cambridge; 31 Sedley Taylor Road, Cambridge. Club: Athenaeum.†

A candidate for an academic appointment is expected to provide his *curriculum vitae* and a list of publications: but what really counts are the reports of his referees. It would not be proper to attempt to assess in public the scientific work of a scholar who is still so active, nor to say anything about his style of thought and technical skill. But the list of published books and papers which is included with this volume would not explain the influence that Mott has had over the thinking of a whole generation of physicists and metallurgists. The historians of science will never know of those timely meetings,

† Reproduced with thanks from *Who's Who*, 1972.

arranged informally to discuss some current scientific problem, out of which sprouted new subjects and new ideas. They cannot catalogue those long sessions of gentle quizzing, where many cherished opinions melt away and new theories formulate themselves. They cannot record the strong questioning voice, teasing simple clarity out of high flown complexity at the conference session, nor can they document the enormous correspondence, hand or typewritten, on all manner of scientific, cultural and educational topics, with all manner of friends, colleagues and acquaintances. They cannot measure the influence of his books and scientific writings, nor the impetus he has given to the writings of others, as an editor of journals and of monograph series.

The contributors to these volumes wish to express, on behalf of all those, from all parts of the world, who have enjoyed his companionship and leadership, as colleagues and as visitors at Bristol and at Cambridge, how much they owe to Nevill Mott – to his friendship, his kindness, and his understanding both in scientific and human affairs, and his encouragement and stimulation in their scientific endeavours.

# PUBLICATIONS OF
# PROFESSOR N. F. MOTT

## BOOKS

*An Outline of Wave Mechanics*, 156 pp. Cambridge University Press, 1930.
*The Theory of Atomic Collisions* (with H. S. W. Massey), 858 pp. Third edition, 1965, Clarendon Press, Oxford, 1934.
*The Theory of the Properties of Metals and Alloys* (with H. Jones), 326 pp. Clarendon Press, Oxford, 1936.
*Electronic Processes in Ionic Crystals* (with R. W. Gurney), 275 pp. Clarendon Press, Oxford, 1940.
*Wave Mechanics and its Applications* (with I. N. Sneddon), 393 pp. Clarendon Press, Oxford, 1948.
*Elements of Wave Mechanics*, 156 pp. Cambridge University Press, 1952.
*Atomic Structure and the Strength of Metals*, 64 pp. Pergamon Press, 1956.
*Electronic Processes in Non-Crystalline Materials* (with E. A. Davis), 437 pp. Clarendon Press, Oxford, 1971.
*Elementary Quantum Mechanics*, 121 pp. Wykeham Publications, London and Winchester, 1972.

## PAPERS

Classical limit of the space distribution law of a gas in a field of force, *Proc. Camb. Phil. Soc.* **24**, 76–9, 1928.
Solution of the wave equation for the scattering of particles by a Coulombian centre of force, *Proc. Roy. Soc.* **A118**, 542–9, 1928.
Scattering of fast electrons by atomic nuclei, *Proc. Roy. Soc.* **A124**, 425–42, 1929.
Interpretation of the relativity wave equation for two electrons, *Proc. Roy. Soc.* **A124**, 422–5, 1929.
Wave mechanics of $\alpha$-ray tracks, *Proc. Roy. Soc.* **A126**, 79–84, 1929.
Quantum theory of electronic scattering by helium, *Proc. Camb. Phil. Soc.* **25**, 304–9, 1929.
Exclusion principle and aperiodic systems, *Proc. Roy. Soc.* **A125**, 222–30, 1929.
Collision between two electrons, *Proc. Roy. Soc.* **A126**, 259–67, 1930.
Scattering of electrons by atoms, *Proc. Roy. Soc.* **A127**, 658–65, 1930.
Influence of radiative forces on the scattering of electrons, *Proc. Camb. Phil. Soc.* **27**, 255–67, 1931.
Theory of effect of resonance levels on artificial disintegration, *Proc. Roy. Soc.* **A133**, 228–40, 1931.
Theory of excitation by collision with heavy particles, *Proc. Camb. Phil. Soc.* **27**, 553–60, 1931.
Polarization of electrons by double scattering, *Proc. Roy. Soc.* **A135**, 429–58, 1932.
Contribution (on the anomalous scattering of fast particles in hydrogen and helium) to the discussion which followed Rutherford's address 'Structure of Atomic Nuclei', *Proc. Roy. Soc.* **A136**, 735–62, 1932.

(With J. M. Jackson.) Energy exchange between inert gas atoms and a solid surface, *Proc. Roy. Soc.* A**137**, 703–17, 1932.

(With H. M. Taylor.) Theory of internal conversion of $\gamma$-rays, *Proc. Roy. Soc.* A**138**, 665–95, 1932.

(With C. D. Ellis.) Internal conversion of $\gamma$-rays and nuclear level systems of thorium B and C, *Proc. Roy. Soc.* A**139**, 369–79, 1933.

(With C. D. Ellis.) Energy relations in the $\beta$-ray type of radioactive disintegration, *Proc. Roy. Soc.* A**141**, 502–11, 1933.

(With H. M. Taylor.) Internal conversion of $\gamma$-rays, Part II, *Proc. Roy. Soc.* A**142**, 215–36, 1933.

(With C. Zener.) Optical properties of metals, *Proc. Camb. Phil. Soc.* **30**, 249–70, 1934.

(With H. Jones and H. W. B. Skinner.) Theory of the form of the X-ray emission bands of metals, *Phys. Rev.* **45**, 379–84. 1934.

Electrical conductivity of metals, *Proc. Phys. Soc. Lond.* **46**, 680–92, 1934.

Resistance of liquid metals, *Proc. Roy. Soc.* A**146**, 465–72, 1934.

Discussion of the transition metals on the basis of quantum mechanics, *Proc. Phys. Soc. Lond.* **47**, 571–88, 1935.

Contribution to 'Discussion on supraconductivity and other low temperature phenomena', *Proc. Roy. Soc.* A**152**, 1–46, 1935.

Electrical conductivity of transition metals, *Proc. Roy. Soc.* A**153**, 699–717, 1936.

Thermal properties of an incompletely degenerate Fermi gas, *Proc. Camb. Phil. Soc.* **32**, 108–11, 1936.

Resistivity of dilute solid solutions, *Proc. Camb. Phil. Soc.* **32**, 281–90, 1936.

(With H. R. Hulme, F. Oppenheimer and H. M. Taylor.) Internal conversion coefficient for $\gamma$-rays, *Proc. Roy. Soc.* A**155**, 315–30, 1936.

Optical constants of copper–nickel alloys, *Phil. Mag.* **22**, 287–90, 1936.

Resistance and thermoelectric properties of the transition metals, *Proc. Roy. Soc.* A**156**, 368–82, 1936.

Energy of the superlattice in $\beta$ brass, *Proc. Phys. Soc. Lond.* **49**, 258–62, Disc. 263, 1937.

Theory of optical constants of Cu–Zn alloys, *Proc. Phys. Soc. Lond.* **49**, 354–6, 1937.

(With H. H. Potter.) Sharpness of the magnetic Curie point, *Nature, Lond.* **139**, 411, 1937.

Cohesive forces in metals, *Sci. Prog. Lond.* **31**, 414–24, 1937.

(With H. Jones.) Electronic specific heat and X-ray absorption of metals and other properties related to electron bands, *Proc. Roy. Soc.* A**162**, 49–62, 1937.

Conduction of electricity in solids, Bristol Conference, Introduction. Also R. W. Gurney and N. F. Mott, Trapped electrons in polar crystals, *Proc. Phys. Soc. Lond.* **49**, 1–2, 32–5, 1937.

(With R. W. Gurney.) Theory of photolysis of silver bromide and photographic latent image, *Proc. Roy. Soc.* A**164**, 151–67, 1938.

Theory of photoconductivity, *Proc. Phys. Soc. Lond.* **50**, 196–200, 1938.

(With M. J. Littleton.) Conduction in polar crystals, Part I. Conduction in polar crystals, Part II. (With R. W. Gurney.) Conduction in polar cyrstals, Part III, *Trans. Faraday Soc.* **34**, 485–511, 1938.

Action of light on photographic emulsions, *Phot. J.* **78**, 286–92, 1938.

Magnetism and the electron theory of metals, *Inst. Physics, Physics in Industry: Magnetism*, 1–15, 1938.

Contribution to discussion of 'Photochemical processes in crystals', by R. Hilsch and R. W. Pohl, *Trans. Faraday Soc.* **34**, 883–8; Disc. 888–92, 1938.

Energy levels in real and ideal crystals, *Trans. Faraday Soc.* **34**, 822–7; Disc. 832–4, 1938.

Absorption of light by crystals, *Proc. Roy. Soc.* A167, 384–91, 1938.

Contribution to discussion on 'Plastic flow in metals', *Proc. Roy. Soc.* A168, 302–17, 1938.

Contact between metal and insulator or semiconductor, *Proc. Camb. Phil. Soc.* **34**, 568–72, 1938.

(With R. W. Gurney.) Recent theories of the liquid state, *Reports on Progress in Physics* **5**, 46–63, 1938.

(With R. W. Gurney.) Theory of liquids, *Trans. Faraday Soc.* **35**, 364–8, 1939.

Faraday Society Discussion on 'Luminescence', *Trans. Faraday Soc.* **35**, 1–240, 1939. R. W. Gurney and N. F. Mott 'Luminescence in solids', 69–73.

Theory of crystal rectifiers, *Proc. Roy. Soc.* A171, 27–38, 1939.

Cu–$Cu_2O$ Photo-cells, *Proc. Roy. Soc.* A171, 281–5, 1939.

(With H. Fröhlich.) Mean free path of electrons in polar crystals, *Proc. Roy. Soc.* A171, 496–504, 1939.

Decomposition of metallic azides, *Proc. Roy. Soc.* A172, 325–34, 1939.

Theory of formation of protective oxide films on metals, *Trans. Faraday Soc.* **35**, 1175–7, 1939.

Reactions in solids, *Rep. prog. Phys.* **6**, 186–211, 1939.

Conference on Internal Strains in Solids, *Proc. Phys. Soc. Lond.* **52**, 1–178, 1940. N. F. Mott and F. R. N. Nabarro, Estimation of degree of precipitation hardening, pp. 86–9.

Theory of formation of protective oxide films on metals, Part II, *Trans. Faraday Soc.* **36**, 472–83, 1940.

Photographic latent image, *Photogr. J.* **81**, 63–9, 1941.

Theorie de l'image latent photographique, *J. Phys. Radium* (8), **7**, 249–52, 1946.

Atomic physics and the strength of metals, *J. Inst. Metals* **72**, 367–80, 1946.

The present position of theoretical physics, *Endeavour* **5**, 107–9, 1946.

(With T. B. Grimley.) The contact between a solid and a liquid electrolyte, *Disc. Faraday Soc.* **1**, 3–11, Disc. 43–50, 1947.

Fragmentation of shell cases, *Proc. Roy. Soc.* A189, 300–8, 1947.

The theory of the formation of protective oxide films on metals, III, *Trans. Faraday Soc.* **43**, 429–34, 1947.

L'oxydation des metaux, *J. de Chimie Physique* **44**, 172, 1947.

Slip at grain boundaries and grain growth in metals, *Proc. Phys. Soc.* **60**, 391–4, 1948.

Notes on latent image theory, *Photogr. J.* **88B**, 119–22, 1948.

(With F. R. N. Nabarro.) Dislocation theory and transient creep, Report of conference on 'Strength of solids' (1947) (*Phys. Soc. Lond.*) 1–19, 1948.

A contribution to the theory of liquid helium II, *Phil. Mag.* **40**, 61–71, 1949.

Theories of the mechanical properties of metals, *Research*, **2**, 162–9, 1949.

The basis of the electron theory of metals, with special reference to the transition metals, *Proc. Phys. Soc. Lond.* A62, 416–22, 1949.

Note on the slowing down of mesons, *Proc. Phys. Soc. Lond.* A62, 196, 1949.

Semiconductors and rectifiers, The 40th Kelvin Lecture, *Proc. Inst. Elec. Eng.* **96**, 253, 1949.

(With N. Cabrera.) Theory of the oxidation of metals, *Rep. Prog. Phys.* **12**, 163–84, 1948–49.

Mechanical properties of metals, *Physica* **15**, 119–34, 1949.

(With Y. Cauchois.) The interpretation of X-ray absorption spectra of solids, *Phil. Mag.* **40**, 1260–9, 1949.

Notes on the transistor and surface states in semiconductors, *Rep. Br. Elec. Res. Ass.* (Ref. L/T 216, 6 pp.), 1949.

(With J. K. Mackenzie.) A note on the theory of melting, *Proc. Phys. Soc. Lond.* A**63**, 411–12, 1950.

Theory of crystal growth, *Nature, Lond.* **165**, 295–7, 1950.

The mechanical properties of metals, *Proc. Phys. Soc. Lond.* B**64**, 729–41, 1951.

Diffusion, work-hardening, recovery and creep, *Solvay Conference, Report*, 1951.

Recent advances in the electron theory of metals, *Prog. Metal Phys.* **3**, 76–114, 1952.

A theory of work-hardening of metal crystals, *Phil. Mag.* **43**, 1151–78, 1952.

The mechanism of work hardening of metals, *39th Thomas Hawksley Lecture; Institution of Mechanical Engineers*, 1952, p. 3.

Dislocations and the theory of solids, *Nature, Lond.* **171**, 234–7, 1953.

Note on the electronic structure of the transition metals, *Phil. Mag.* **44**, 187-90, 1953.

A theory of work-hardening of metals, II. Flow without slip-lines, recovery and creep, *Phil. Mag.* **44**, 742–65, 1953.

Ricerche recenti in teoria dei solidi, Suppl. *Nuovo Cim.* **10**, 212–24, 1953.

Dislocations, plastic flow and creep (Bakerian Lecture), *Proc. Roy. Soc.* A**220**, 1–14, 1953.

Difficulties in the theory of dislocations, *Proc. Internat. Conf. Theoretical Phys., Kyoto and Tokyo*, Sept. 1953, pp. 565–70.

Creep in metals: the rate determining process, Paper No. 2 of *Symposium on Creep and Fracture of Metals at High Temperatures*, 31 May, 1 and 2 June, 1954, National Physical Laboratory publication, pp. 1–4.

Science and modern languages, *Mod. Lang.* **36**, 45–50, 1955.

A theory of fracture and fatigue, *J. Phys. Soc. Jap.* **10**, 650–6, 1955.

Physics of the solid state, *Advancement of Science*, no. 46, Sept. 1955, pp. 1–9.

*Science and education (Thirty-sixth Earl Grey Memorial Lecture).* Delivered at King's College, Newcastle upon Tyne, 11 May 1956, printed by Andrew Reid and Co., Newcastle upon Tyne, pp. 1–15.

Creep in metal crystals at very low temperatures, *Phil. Mag.* **1**, 568–72, 1956.

On the transition to metallic conduction in semiconductors, *Can. J. Phys.* **34**, 1356–68, 1956.

Fracture in metals, *J. Iron Steel Inst.* **183**, 233–43, 1956.

Theoretical chemistry of metals, *Nature, Lond.* **178**, 1205–7, 1956.

A discussion on work-hardening and fatigue in metals, *Proc. Roy. Soc.* A**242**, 145–7, 1957.

(With J. W. Mitchell.) The nature and formation of the photographic latent image, *Phil. Mag.* **2**, 1149–70, 1957.

(With K. W. H. Stevens.) The band structure of the transition metals, *Phil. Mag.* **2**, 1364–86, 1957.

The physics and chemistry of metals, *Year Book of the Physical Society* 1957. pp. 1–13.

A theory of the origin of fatigue cracks, *Acta Met.* **6**, 195–7, 1958.

The transition from the metallic to the non-metallic state. Suppl., *Nuovo Cim.* **7**, 312–28, 1958.

(With T. P. Hoar.) A mechanism for the formation of porous anodic oxide films on aluminium, *J. Phys. Chem. Sol.* **9**, 97–9, 1959.

La rupture des metaux (*J. Iron Steel Inst.* **183**, 233–43, 1956). Paris: Imprimerie Nationale, 1959.

The work-hardening of metals, *Trans. Met. Soc. of AIME* **218**, 962–8, 1960.

(With R. J. Watts-Tobin.) The interface between a metal and an electrolyte, *Electrochim. Acta* **3**, 79–107, 1961.

The transition to the metallic state, *Phil. Mag.* **6**, 287–309, 1961.

(With W. D. Twose.) The theory of impurity conduction, *Phil. Mag. Suppl.* (*Adv. Phys.*), **10**, 107–63, 1961.

(With R. Parsons, and R. J. Watts-Tobin.) The capacity of a mercury electrode in electrolytic solution, *Phil. Mag.* **7**, 483–93, 1962.

The cohesive forces in metals and alloys, *Rep. Prog. Phys.* **25**, 218–43, 1962.

Opening address to the International Conference on the Physics of Semiconductors held at Exeter, July 1962, The Institute of Physics and The Physical Society, 1–4, 1962.

The structure of Metallic Solid Solutions, *J. Phys. Radium* **23**, 594–6, 1962.

Atomic physics and the strength of metals. The Rutherford Memorial Lecture, 1962, *Proc. Roy. Soc.* A**275**, 149–60, 1963.

Physics in the Modern World, *Proc. Ghana Acad. Sci.* **2**, 80–4, 1964.

Electrons in transition metals, *Adv. Phys.* **13**, 325–422, 1964.

On teaching quantum phenomena, *Contemp. Phys.* **5**, 401–18, 1964.

The theory of magnetism in transition metals, *Proceedings of the International Conference on Magnetism, Nottingham*, 7–11 September 1964, pp. 67–8.

An outline of the theory of transport properties, '*Liquids: Structure, Properties, Solid Interactions*, pp. 152–71. Edited by T. J. Hughel, Elsevier Publishing Company, Amsterdam, 1965.

The electrical properties of liquid mercury, *Phil. Mag.* **13**, 989–1014, 1966.

Electrons in disordered structures, *Adv. Phys.* **16**, 49–144, 1967.

(With R. S. Allgaier.) Localized states in disordered lattices, *Phys. Stat. Sol.* **21**, 343, 1967.

The Solid State, *Scientific Am.* **217**, 80, 1967.

The transition from metal to insulator, *Endeavour* **26**, 155–8, 1967.

Conduction in non-crystalline systems, I. Localized electronic states in disordered systems, *Phil. Mag.* **17**, 1259–68, 1968.

(With E. A. Davis.) Conduction in non-crystalline systems, II. The metal–insulator transition in a random array of centres, *Phil. Mag.* **17**, 1269–84, 1968.

Conduction in glasses containing transition metal ions, *J. Non-Crystalline Solids* **1**, 1–17, 1968.

Metal–insulator transition, *Reviews of Modern Physics* **40**, 677–83, 1968.

Polarons in transitional-metal oxides, *Comments on Solid State Physics* **1**, 105–11, 1968.

Conduction in non-crystalline materials, III. Localized states in a pseudogap and near extremities of conduction and valence bands, *Phil. Mag.* **19**, 835–52, 1969.

Conduction and switching in non-crystalline materials, *Contemporary Physics* **10**, 125–38, 1969.

(With I. G. Austin.) Polarons in crystalline and non-crystalline materials, *Adv. Phys.* **18**, 41–102, 1969.

Charge transport in non-crystalline semiconductors, *Festkörperprobleme* **9**, 22–45, 1969.

Metallic and non-metallic behaviour in compounds of transition metals, *Phil. Mag.* **20**, 1–21, 1969.

(With M. Cutler.) Observation of Anderson localization in an electron gas, *The Physical Review* **181**, 1336–40, 1969.

Conduction in solids, *Nuovo Cimento* **1**, 174–83, 1969.

(With R. Catterall.) Metal–ammonia solutions, *Advances in Physics*, **18**, 665–80, 1969.

(With F. P. Fehlner.) Low-temperature oxidation, *Oxidation of Metals* **2**, 59–99, 1970.

(With Z. Zinamon.) Metal–non-metal transitions in narrow band materials; crystal structure versus correlation, *Phil. Mag.* **21**, 881–95, 1970.

(With I. G. Austin.) Metallic and nonmetallic behavior in transition metal oxides, *Science* **168**, 78–83, 1970.

Conduction in non-crystalline systems, IV. Anderson localization in a disordered lattice, *Phil. Mag.* **22**, 7–29, 1970.

The metal–non-metal transition, *Comments on Solid State Physics* **2**, 183–92, 1970.

(With E. A. Davis.) Conduction in non-crystalline systems. V. Conductivity, optical absorption and photoconductivity in amorphous semiconductors, 1970, *Phil. Mag.* **22**, 903–22, 1970.

(With Z. Zinamon.) The metal–nonmetal transition, *Reports on Progress in Physics*, **33**, 881–940, 1970.

Unsolved problems in the theory of conduction in noncrystalline materials, *Comments on Solid State Physics* **3**, 123–7, 1970.

Conduction in glassy and liquid semiconductors, *Discussions of The Faraday Society*, No. 50, 7–12, 1970.

La théorie des métaux de 1930 à 1970, *Mémoires Scientifiques Rev. Métallurg.* **68**, 125–30, 1971.

Conduction in non-crystalline systems. VI. Liquid semiconductors, *Phil. Mag.* **24**, 1–18, 1971.

The metal–non-metal transition, *Journal de Physique*, Colloque No. *C*1, Suppl. **32**, *C*11–14, 1971.

(With J. V. Acrivos.) On the metal–non-metal transition in sodium-ammonia solutions, *Phil. Mag.* **24**, 19–31, 1971.

(With R. M. White.) The metal–non-metal transition in nickel sulphide (NiS), *Phil. Mag.* **24**, 854–6, 1971.

Conduction in non-crystalline systems, VII. Non-ohmic behaviour and switching, *Phil. Mag.* **24**, 911–34, 1971.

Conduction in non-crystalline systems, VIII. The highly correlated electron gas in doped semiconductors and in vanadium monoxide, *Phil. Mag.* **24**, 953–8, 1971.

(With L. Friedman.) The Hall effect near the metal–insulator transition, *J. Non-Crystalline Solids* **7**, 103–8, 1972.

Introductory talk; condution in non-crystalline materials, *J. Non-Crystalline Solids* **8–10**, 1–18, 1972.

Evidence for a pseudogap in liquid mercury, *Phil. Mag.* **26**, 505–22, 1972.

Conduction in non-crystalline systems, IX. The minimum metallic conductivity, *Phil. Mag.* **26**, 1015–26, 1972.

Chemistry and Physics; A discussion of the role of both in our understanding of glass, *Bulletin de la Société Chimique Beograd* **37**, 71–83, 1972.

The electrical resistivity of liquid transition metals, *Phil. Mag.* **26**, 1249–61, 1972.

The metal–insulator transition in extrinsic semiconductors, *Adv. in Phys.* **21**, 785–823, 1972.

Properties of compounds of type $Na_xSi_{46}$ and $Na_xSi_{136}$, *J. Solid State Chemistry* **6**, 348–51, 1973.

Transport in disordered materials, *Phys. Rev. Lett.* **31**, 466–7, 1973.

Etude comparative du comportement magnetique des phases $LaNiO_3$ et $LaCuO_3$, *Mat. Res. Bull.* **8**, 647–56, J. B. Goodenough, N. F. Mott, M. Pouchard, G. Demazeau and P. Hagenmuller, 1973.

Conduction in amorphous materials, *Electronics and Power* **19**, 321–4, 1973.

Some problems about polarons in transition metal compounds, *Cooperative Phenomena* edited by H. Haken and M. Wagner, Springer-Verlag, pp. 2–14, 1973.

Metal–insulator transition, *Comtemp. Phys.* **14**, 401–13, 1973.

# CHAPTER 1

# POINT DEFECTS

## *by* J. D. ESHELBY †

## 1.1 INTRODUCTION

In this chapter we shall mainly be concerned with point defects
(vacancies and interstitials) in pure crystals, though substitutional
and interstitial impurities will receive some attention, particularly
as ideas originally developed to deal with them have subsequently
been applied to vacancies and self-interstitials.

Section 1.2 is concerned with the formation of defects by thermal
excitation, cold work and irradiation. Part of section 1.3 (on defect
mobility) may appear a little out of proportion. However, it has
seemed to the writer that many solid state and metallurgical texts
introduce the standard jump frequency formula with much less discussion than they devote to related matters, and that therefore a
slightly extended treatment at an intermediate level might be useful.

Section 1.4 (Physical properties of defects) is concerned with some
of the basic physical properties (energy and volume of formation,
effect on electrical resistance) of defects and their measurement.
Since the presence of defects has usually to be inferrred from their
influence on bulk properties this section could equally well be entitled
'Effect of defects on physical properties'. It is perhaps best to confine
this latter phrase to secondary effects consequent on the basic effects
mentioned above. Important among them is the interaction between
point defects and dislocations. Section 1.5 ('Interaction energies of
point defects') serves as a link with other chapters.

## 1.2 TYPES OF DEFECT. THEIR FORMATION BY THERMAL ACTIVATION, IRRADIATION AND PLASTIC DEFORMATION

Starting from a collection of simple lattice vacancies one can form
more elaborate defects by allowing them to aggregate into clusters.

† Dr Eshelby is Professor at the Department of the Theory of Materials, University
of Sheffield.

A pair of vacancies on neighbouring sites constitute a divacancy, a dumb-bell shaped configuration which can assume any one of $\frac{1}{2}z$ orientations in a lattice with co-ordination number $z$. The stable configuration of a more complicated cluster is less obvious. One might guess that a trivacancy in a face-centred cubic lattice would take the form of a triangle of vacancies lying in a $\{111\}$ plane. However, detailed calculations suggest that the energy will be lowered if a suitable adjacent atom moves in towards the triangle, leaving a configuration which may be described as a tetrahedron of vacancies

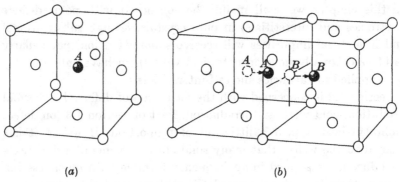

(a)                                        (b)

Fig. 1.1 Body-centred (a) and split (b) interstitial in a face-centred cubic lattice.
(Damask and Dienes, 1963)

with an atom in the middle (Damask *et al.*, 1959). Computer calculations of the equilibrium configuration for various clusters of vacancies (and interstitials) have been made by Vineyard (1961) and his associates.

In contrast to the vacancy even a single interstitial can display quite complicated behaviour. Fig. 1.1(a) shows an interstitial atom $A$ in a body-centred position in a face-centred cubic lattice. In fig. 1.1(b) the atom has moved to the right, displacing the atom $B$, so that in the resulting dumb-bell configuration one cannot tell whether $A$ or $B$ is 'really' the interstitial. The resulting configuration is called a split interstitial (German *Hantel*). Calculations indicate that its energy is slightly less than that of the simple interstitial of fig. 1.1(a). There is, in fact, experimental evidence that split interstitials can exist in several metals. If the atom $A$ in fig. 1.1(a) had been displaced vertically we should have got a vertical dumb-bell, and a displacement out of the paper would have given a horizontal dumb-bell, perpendicular to the one in fig. 1.1(b). The defect can thus exist in one of three orientations, and can change from one to another. There will evidently be a tetragonal distortion of the lattice near the defect, with its tetragonal axis in the [100], [010] or [001] direction, according to the orientation of the defect. This is very similar to what happens with interstitial carbon (or nitrogen) atoms in body-centred iron. Here the carbon atom can lie at the mid-point of a line joining two iron atoms and oriented in either the [100], [010] or [001] direction. Relaxation due to reorientation under the influence of applied stresses leads to a peak in the internal friction spectrum (Snoek effect), and interaction with Bloch walls affects the magnetic susceptibility. We might expect split interstitials to produce similar effects, and they have in fact been found in nickel and perhaps copper, into which interstitials have been introduced by plastic deformation or irradiation (Seeger and Wagner, 1965).

Point defects can be introduced into a crystal by heating it (thermal excitation) deforming it plastically, or irradiating it with fast particles.

At a finite temperature an otherwise perfect crystal will contain point defects of the types mentioned in section 1.1, and others, in concentrations which are determined by their energies and entropies of formation.

In order to find the equilibrium concentration of a particular type of defect we may start from a result of statistical mechanics which says that the probability of finding a system in a definite state whose Gibbs free energy is $G$ is given by

$$p = C \exp\left(-G/kT\right)$$

where $k$ is Boltzmann's constant, $T$ the absolute temperature and $C$ is a normalising constant. With a suitable choice of $C$, $G$ may be taken to be the excess of the Gibbs free energy over its value in some standard state. $G$ is then the work required to carry the system from the standard to the actual state. If the system is subject to an external pressure, or some more complicated applied stress, part of this work goes towards increasing the potential energy of the loading mechanism. Use of the Gibbs free energy $G$, rather than the Helmholtz free energy $F$, automatically allows for this, For a crystal under atmospheric pressure the difference between $G$ and $F$ is negligible.

For definiteness we consider the case of vacancies. Let the standard state be a crystal free of vacancies, and let $\Delta G$ be the work required to form one vacancy. The work required to form $n$ vacancies will then be $G = n\Delta G$ if we can neglect the interaction between them. The probability of finding $n$ vacancies at $n$ specified sites is thus $C\exp(-n\Delta G/kT)$. The probability of finding $n$ vacancies at $n$ unspecified sites in a crystal containing $N$ atoms (and hence $N+n$ lattice sites) is thus

$$p(n) = C \frac{(N+n)!}{N!\,n!} e^{-n\Delta G/kT} \qquad (1.1)$$

where the collection of factorials is the number of distinct ways of picking out $n$ unoccupied sites from the $N+n$ (occupied or unoccupied) sites. The most probable value of $n$ is that for which a small change of $n$ leaves $p(n)$ nearly unaltered, so that, say, $p(n) = p(n-1)$. After extensive cancellation this gives

$$n = (N+n)\,e^{-\Delta G/kT} \qquad (1.2)$$

or a most probable concentration (vacant sites divided by total sites)

$$c = e^{-\Delta G/kT}. \qquad (1.3)$$

One can show that (1.2) also gives, very closely, the average value of $n$,

$$\bar{n} = \frac{\sum_n n p(n)}{\sum_n p(n)} = \frac{\sum_n n[(N+n)!/N!\,n!]\,x^n}{\sum_n[(N+n)!/N!\,n!]\,x^n}, \quad \text{where } x = e^{-\Delta G/kT}. \quad (1.4)$$

In the summations, terms for which $n$ is not small in comparison with $N$ are inaccurate because of interaction between the defects, or even meaningless (more vacancies than atoms), but they have a negligible effect on the sum for small $x$ and we may carry the summations from 0 to $\infty$. The denominator in (1.4) is then recognisable as the expansion of $(1-x)^{-(N+1)}$ by the binomial theorem. If we differentiate the denominator with respect to $x$ we get $x^{-1}$ times the numerator. Consequently

$$\bar{n} = \frac{x\,d(1-x)^{-(N+1)}/dx}{(1-x)^{-(N+1)}} = (N+1)\frac{x}{1-x}$$

which agrees with (1.2) for large $N$ and small $x$.

The number of interstitials present in thermal equilibrium may be calculated in the same way. The combinatorial factor in (1.1) may be more complicated if the number of interstitial sites differs from the number of lattice sites, but, as might be expected, the ratio of occupied to total interstitial sites is still given by (1.3) with a suitable $\Delta G$.

More complex defects, in particular clusters of vacancies, may be handled in the same way. For example, a divacancy may be regarded as a dumb-bell which may exist in a number of orientations. Each orientation may be regarded as a separate type of defect and its concentration calculated from (1.3). Multiplication of the number of orientations then gives the total number of defects. However, to get the combinatorial factor in (1.1) right we must take care to choose the number of independent orientations in such a way that two pairs apparently having different orientations and different positions are not in fact the same physical pair. We illustrate this for divacancies in a lattice with arbitrary co-ordination number, and for two types of trivacancy in a face-centred cubic lattice.

An observer looking at the crystal will see divacancies situated at various points in the crystal and with different orientations. If he is not looking along a symmetry element of the crystal lattice he can classify the two vacancies of a pair into a near one and a far one. A particular near vacancy can have its companion in any one of $\frac{1}{2}z$ positions, where $z$ is the number of nearest neighbour sites; if the far vacancy were in one of the remaining $\frac{1}{2}z$ positions it would be the near one. For a particular one of these $\frac{1}{2}z$ orientations the near vacancy can lie at any one of the $N$ lattice sites, so the concentration of them is given by (1.3) with a suitable $\Delta G$. There will be an equal concentration of divacancies with any one of the remaining $\frac{1}{2}z - 1$ orientations, so the concentration of divacancies is

$$c = \tfrac{1}{2}z\,\mathrm{e}^{-\Delta G_2/kT}.$$

The most obvious form for a trivacancy in a face-centred cubic lattice is an equilateral triangle of vacancies lying in any one of the four sets of {111} planes. An observer can classify them as looking like $\triangle$ or $\triangledown$, and can distinguish a 'near' atom in the triangle. The near atom in a $\triangle$ can be situated at any of the $N$ atom positions in the lattice, so that the number of them is given by (1.3) with a suitable $\Delta G$. There is an equal number of $\triangledown$s in (111) planes, or in all

$$n = 2N\,\mathrm{e}^{-\Delta G_3/kT} \tag{1.5}$$

trivacancies in (111) planes. There are equal numbers lying in the other three {111} planes so that the total concentration of trivacancies is

$$c_3 = 8\,e^{-\Delta G_3/kT}. \qquad (1.6)$$

However, calculation (Vineyard, 1961) suggests that a trivacancy in the form of a tetrahedron of vacancies with an atom in the middle will have an energy of formation, $\Delta G_3'$ say, less than $\Delta G_3$. Any tetrahedron will have one face in a (111) plane and it may be a $\triangle$ or a $\triangledown$. If the (111) planes are labelled $ABC\ldots$ in the usual way, a $\triangle$ in an $A$ plane can only accept the fourth vacancy in the $B$ position and a $\triangledown$ in an $A$ plane can only accept one in the $C$ position, so that there is only one type of tetrahedron associated with a $\triangle$ and one with a $\triangledown$. The whole number of tetrahedra in the crystal is thus the same as the number of triangular trivacancies which happen to lie on (111) planes, equation (1.5), apart from the change in $\Delta G$. So

$$c_3' = 2\,e^{-\Delta G_3'/kT} \qquad (1.7)$$

is the concentration of tetrahedral trivacancies. Of course the two types co-exist in the proportion corresponding to (1.6), (1.7), but if $\Delta G_3'$ is notably smaller than $\Delta G_3$ the tetrahedral type will predominate at reasonable temperatures and may be regarded as 'the' trivacancy.

It is customary to write an expression like (1.3) for a defect concentration in terms of the energy and entropy of formation rather than in terms of $\Delta G$. It will be convenient to consider this in section 1.4.2.

Theory and experiment both indicate that the energy required to form an interstitial is much greater than the energy required to form a vacancy, so that practically speaking only vacancies are formed thermally. Typically about 1 lattice site in $10^4$ may be vacant in a metal just below the melting point.

If a metal is bombarded with high-energy radiation (neutrons, protons, deuterons, $\alpha$-particles, electrons) atoms are displaced from their lattice sites and ultimately take up interstitial positions, not necessarily near their point of origin. Before settling down they may knock out other atoms, which in turn may displace others. In this way, in contrast to thermal generation, equal numbers of vacancies and interstitials are formed. It is interesting to see how the concentration produced by a given dose of irradiation compares with the number generated thermally. The following is a rather crude estimate for the case of copper irradiated by neutrons.

If the energy of the incident neutron, $E_n$ say, greatly exceeds the energy $E_d$ required to displace the struck atom, the latter can be treated as free. Elementary mechanics then gives for the maximum energy transferred to the atom the value $E_m = 4E_n M_1 M_2/(M_1 + M_2)^2$ where $M_1$, $M_2$ are the masses of the colliding particles, or say $4E_n/A$ for a neutron colliding with a particle of atomic weight $A$. The average energy transferred is $\frac{1}{2}E_m$.

The number of collisions per cm³ produced by a flux of $\phi$ neutrons per cm² maintained for a time $t$ is $n = \phi t \sigma n_0$, where $n_0$ is the number of atoms per cm³ and $\sigma$ is the collision cross-section. The displaced particle will have an average energy $\frac{1}{2}E_m$ and should thus be able to create about $\frac{1}{2}E_m/E_d$ additional Frenkel pairs; a somewhat more accurate estimate is $\frac{1}{2}E_m/2E_d$. Hence finally the total concentration of pairs is

$$c = \frac{n}{n_0}\frac{E_m}{4E_d} = \frac{\sigma E_n}{A E_d}\,\phi t.$$

For copper $A = 64$, $E_d \sim 25$ eV and for 1 MeV neutrons $\sigma \sim 1$ barn $= 10^{-24}$ cm², so that $c \sim 6 \times 10^{-22}\,\phi t$, and a day in a typical pile flux of $10^{13}$ neutrons cm⁻² sec⁻¹ can give a pair concentration $6 \times 10^{-4}$, comparable with the number of thermal vacancies just below the melting point.

The fact that the atoms are regularly arranged leads to some interesting effects in irradiated materials. If the energy transfer is low a succession of collisions may occur along a close-packed row of atoms. Each atom returns to its equilibrium position, so that energy, but not material, is transported along the chain (focuson). If the conditions of the initiating collision are somewhat different a vacancy may be formed in the chain, and consequently somewhat further along it $n + 1$ atoms must be distributed over the distance normally occupied by $n$. Such a 'crowdion' configuration (essentially Frenkel and Kontorova's (1938) 'caterpillar') is formally a small loop of edge dislocation with its Burgers vector along the direction of the row, so that its movement transports material along the row, though no individual atom moves a large distance. Individual atoms may travel large distances along channels in the crystal structure.

In the case of plastic deformation, point defects probably mostly arise from the non-conservative motion of jogs on dislocations, produced by their intersection with other dislocations. Seitz (1952) has given the estimate $c \simeq 10^{-4}\,\epsilon$ for the concentration produced by a plastic strain $\epsilon$.

## 1.3 THE MOBILITY OF POINT DEFECTS

Under the influence of thermal vibrations an interstitial impurity may jump from one interstitial site to the next. An atom adjacent to a vacancy may fall into it; the vacancy has then moved to a new position. These are simple diffusion processes. More complex re-arrangements can occur – for example a divacancy or a split inter-stitial or an interstitial carbon atom in alpha-iron may re-orientate.

It is a problem in kinetic theory (or the theory of rate processes) to find the rate at which such rearrangements occur. It is convenient to start with the highly idealised situation shown in fig. 1.2. There is a

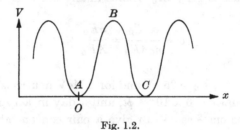

Fig. 1.2.

particle in the well $A$ of the one-dimensional potential $V(x)$. It is required to find the probability that, as a result of thermal agitation, the particle will leave the well by passing over the barrier $B$. It is convenient to imagine that there is a large number of non-interacting particles in the well as we can then discuss the decrease in the number of particles in $A$ rather than the decrease in the probability that the single particle is still there. This is of course a purely formal device. Fig. 1.2 may be taken as referring to a mental superposition of a large number of wells each containing one or no particles. The collection may refer to a number of actual wells at different points in a crystal, or it may just be an 'ensemble' in the sense of statistical mechanics.

With this understanding we may say that there are $n$ non-inter-acting particles in $A$, and that as a result of thermal agitation a number of them will jump over the barrier $B$. (As many will jump out in the opposite direction, but we are not concerned with them.) It is convenient to specify the rate at which the particles leave by way of a jump frequency $f$ defined by

$$f = -\frac{1}{n} \cdot \frac{1}{2} \cdot \frac{dn}{dt}.$$

Its value is

$$f = \nu \, e^{-U/kT} \tag{1.8}$$

where $U$ is the height of the barrier and $\nu$ is the frequency with which a particle vibrates in the bottom of the well. The justification usually given for (1.8) is that the exponential gives the probability that a particle has enough energy to surmount the barrier and that $\nu$ is the number of attempts it makes in unit time. Though it gives the correct result this argument is, to say the least, rather unconvincing. Since (1.8) and generalisation of it to more realistic situations than the one shown in fig. 1.2 play such an important part in the theory of defect mobility and diffusion it is perhaps worth devoting some space to the matter. We shall give a simple derivation of (1.8) and extend it to motion in a fixed two- or three-dimensional potential. Finally, following Vineyard (1957) we indicate how the method can in principle be extended to the realistic case where we cannot regard the particle as moving in a fixed potential (because the atoms which produce the potential are themselves displaced by the motion of the particle) or where (as in the motion of a vacancy or the re-orientation of a divacancy) the configuration of the defect cannot be described by giving the co-ordinates of a single particle.

To calculate the rate at which particles leave the well $A$ by passing over the barrier $B$ from left to right we only need two results from statistical mechanics. First, Boltzmann's barometric formula states that the number of particles per unit length of the $x$-axis is const. $\exp[-V(x)/kT]$. Secondly Maxwell's distribution law for velocities states that the average velocity is a constant independent of $V(x)$. This applies to any average velocity of those particles which are moving from left to right, $\bar{v}$, say. (Actually any average velocity is a multiple of $\sqrt{(kT/m)}$ for a particle of mass $m$, but we do not need to know this.)

Define the flux of particles at $x$ to be the number of particles which in unit time pass the point $x$ moving from left to right. It is equal to the number of particles per unit length at $x$ multiplied by $\bar{v}(x)$. Then we have

$$\frac{\text{flux at } B}{\text{flux at } A} = \frac{\bar{v}(B)\exp[-V(B)/kT]}{\bar{v}(A)\exp[-V(A)/kT]}.$$

The $\bar{v}$s cancel, since, as we have seen, $\bar{v}$ is in fact independent of position, and so with $U = V(B) - V(A)$ we have

$$\text{flux at } B = \text{flux at } A \times \exp(-U/kT).$$

The flux at $A$ can be calculated directly. If $\nu$ is the frequency of vibration of a particle in the potential well, each of the $n$ particles at the bottom of the well passes the centre line of the well from left to right $\nu$ times a second, and the flux is $n\nu$. So

$$\text{flux at } B = n\nu \exp(-U/kT)$$

and consequently for the jump frequency we have

$$f = \frac{\text{flux at } B}{\text{number in } A} = \nu \exp(-U/kT)$$

in agreement with (1.8).

The artifice which gave us the flux at $A$ becomes rather hard to apply in two or more dimensions, so we re-derive (1.8) by a method which lends itself to extension.

According to Maxwell and Boltzmann the number of particles between $x$ and $x + dx$ having momentum between $p$ and $p + dp$ is $N(x,p)\,dx\,dp$ where

$$N(x,p) = \text{const.} \exp\{[-V(x) - p^2/2m]/kT\}$$

where $m$ is the mass of a particle. The flux at $B$ is $N(x,p)$ evaluated at $B$, multiplied by the velocity $p/m$ and integrated with respect to $p$ from 0 to $\infty$, since we are only interested in particles moving from left to right. The number of particles in the well $A$ is $N(x,p)$ integrated over all momenta and over a range of $x$ which embraces all the particles which are considered to be in the well. Hence $f$, the ratio of flux to the number of particles in the well, is given by

$$f = \frac{e^{-U/kT}\int_0^\infty (p/m)\,e^{-p^2/2mkT}\,dp}{\int e^{-V(x)/kT}\,dx \int_{-\infty}^\infty e^{-p^2/2mkT}\,dp}.$$

We now suppose that $kT \ll U$. The density of particles is then negligible except at the bottom of the well and we may without noticeable error take the limits of the $x$-integral in the denominator to be $\pm\infty$. (It is only in such a case that one can make a sensible distinction between particles in the well and particles in transit to the next well.) For the same reason we may replace $V(x)$ by the first non-vanishing term of its Taylor expansion. If $A$ is at $x = 0$ this is

$$V(x) = \tfrac{1}{2} V''(0) x^2$$

because $V'$ is zero at a minimum and we have chosen $V(0)$ to be zero.

The integrals can be evaluated with the help of

$$\int_0^\infty e^{-t^2/a^2} \, \mathrm{d}t = \tfrac{1}{2} a \sqrt{\pi}, \quad \int_0^\infty t \, e^{-t^2/a^2} \, \mathrm{d}t = \tfrac{1}{2} a \qquad (1.9)$$

to give (1.8) with

$$\nu = \frac{1}{2\pi} \sqrt{\left( \frac{V''(0)}{m} \right)} \qquad (1.10)$$

which is precisely the frequency of a particle of mass $m$ vibrating in a potential well with curvature $V''(0)$.

We consider next the two-dimensional situation of fig. 1.3. There are $n$ particles in the well $A$ and it is required to find the rate at which they pass through the saddle-point region $B$ on the way to the well $C$. Near $A$ we may write the potential as

$$V = \tfrac{1}{2} V_x'' x^2 + \tfrac{1}{2} V_y'' y^2 \qquad (1.11)$$

where $x$ and $y$ are axes chosen parallel to the principal axes of the elliptical equipotentials around $A$, and near $B$ we may put

$$V = U - \tfrac{1}{2} |V_{x'}''| x'^2 + \tfrac{1}{2} V_{y'}'' y'^2 \qquad (1.12)$$

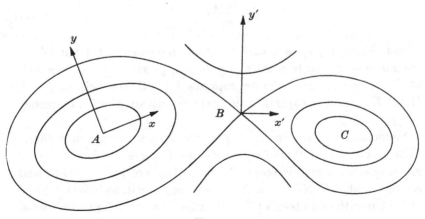

Fig. 1.3.

where $x'$, $y'$ are local axes at $B$ parallel to the principle axes of the hyperbolic equipotentials near $B$. $V''_x$ stands for $(\partial^2 V/\partial x^2)_A$, and similarly for the others. The coefficient of $x'^2$ is negative because, along $x'$, $V$ decreases on either side of $B$.

The two-dimensional analogue of $N(x,p)$ is given by

$$N(x,y,p_x,p_y)\,\mathrm{d}x\,\mathrm{d}y\,\mathrm{d}p_x\,\mathrm{d}p_y$$
$$= \text{const.}\exp\{[-V(x,y)-(p_x^2+p_y^2)/2m]/kT\}\,\mathrm{d}x\,\mathrm{d}y\,\mathrm{d}p_x\,\mathrm{d}p_y. \quad (1.13)$$

A particle may be said to have crossed the saddle-point when it has passed over the line $x' = 0$ from left to right. The flux of particles through the saddle-point is thus equal to $N$ (equation (1.13)) multiplied by the velocity $p_{x'}/m$ and integrated along the $y'$ axis and over positive $p_{x'}$. The number of particles in $A$ is $N$ integrated over all momenta and over the area of the well. We may use the same argument as in the one-dimensional case to replace $V$ by the expressions (1.11) and (1.12) and to extend the spatial integrations between $\pm\infty$. The jump frequency, the ratio of flux at $B$ to particles in $A$, thus takes the form

$$f = e^{-U/kT}$$

$$\times \frac{\displaystyle\int_{-\infty}^{\infty} e^{-V''_{y'}y'^2/2kT}\,\mathrm{d}y' \int_0^{\infty}(p_{x'}/m)\,e^{-p_{x'}^2/2mkT}\,\mathrm{d}p_{x'}}{\displaystyle\int_{-\infty}^{\infty} e^{-V''_x x^2/2kT}\,\mathrm{d}x \int_{-\infty}^{\infty} e^{-V''_y y^2/2kT}\,\mathrm{d}y \int_{-\infty}^{\infty} e^{-p_x^2/2mkT}\,\mathrm{d}p_x \int_{-\infty}^{\infty} e^{-p_y^2/2mkT}\,\mathrm{d}p_y}.$$
$$(1.14)$$

Using (1.9) this becomes

$$f = \frac{\nu_x \nu_y}{\nu_{y'}} e^{-U/kT} \quad (1.15)$$

where the $\nu_x$, $\nu_y$, $\nu_{y'}$ are the vibration frequencies of a particle of mass $m$ constrained to move along the $x$ or $y$ axis at $A$, or the $y'$ axis at $B$, calculated from the appropriate $V''$ with the help of (1.11), (1.12). Equation (1.15) is the same as (1.8) with an effective frequency $\nu_x \nu_y / \nu_{y'}$.

Suppose next that fig. 1.3 represents a cross-section of a three-dimensional potential field with a well at $A$ where the equipotentials are ellipsoids, a saddle-point $B$ where they are hyperboloids, and another well at $C$. Near $A$ and $B$ the potential will be like (1.11) and (1.12) but with extra terms $\frac{1}{2}V''_z$, $\frac{1}{2}V''_{z'}$ where $z$ and $z'$ refer to additional co-ordinate axes erected perpendicular to the paper at $A$ and $B$.

The distribution of particles over position and momentum is given by an obvious generalisation of (1.13).

The jump frequency is now the flux of particles from left to right across the $y'z'$ plane at $B$, divided by the number of particles in the well at $A$. It is easy to see how (1.14) is to be modified. In the numerator (flux) the velocity integral over $p_{x'}$ is unaltered, but since we are integrating over the $y'z'$ plane we must insert an integral over $z'$ similar to the one over $y'$. In the denominator (number of particles in $A$) we must insert a $z$ integral similar to the $x$ and $y$ integrals and a $p_z$ integral similar to the $p_x$ and $p_y$ integrals. Applying (1.9) and (1.10) will evidently give

$$f = \frac{\nu_x \nu_y \nu_z}{\nu_{y'} \nu_{z'}} \, \mathrm{e}^{-U/kT} \qquad (1.16)$$

with an obvious generalisation of notation.

The foregoing calculations are unrealistic because we are actually interested in the motion of a particle which is moving in a potential which is varying in time because the atoms which produce it are subject to thermal fluctuations and are themselves affected by the movement of the atom we are interested in. Moreover, if the jump process is a complicated one (e.g. the movement of a vacancy, the re-orientation of a split interstitial) it may not be clear how to introduce a co-ordinate which describes its progress. These difficulties are in principle avoided by a treatment on the following lines.

Consider a crystal containing $N$ atoms, not necessarily all the same. Its total potential energy will be a function of the $3N$ variables $x_n, y_n, z_n, n = 1, 2, \ldots N$. Imagine that $x_1, y_1, \ldots z_N$ are taken as $3N$ rectangular co-ordinates in a $3N$-dimensional 'configuration space' and that contours $V = $ const. are plotted in it. There will be a number of potential wells in it, representing for example, the perfect crystal, the crystal containing a vacancy at a specific site, the crystal containing two vacancies at two specific sites, the crystal containing an interstitial at a specific site, and so on. Between the minima there will be saddle-points. We may take fig. 1.3 as a crude picture of configuration space; the place of the moving particle is taken by a representative point whose movement is dictated by the way the co-ordinates $x_1, y_1, \ldots z_N$ change with time.

A representative point oscillating near the bottom of well $A$ will in general execute a $3N$-dimensional Lissajous figure. However, if it is vibrating along one of the $3N$ principal axes of the hyper-ellipsoidal potential surfaces around $A$, its motion is simple harmonic. If we rotate our axes to coincide with these directions we get a set of so-called normal co-ordinates. The associated frequencies $\nu_i$, $i = 1, 2, \ldots 3N$ are the possible vibration frequencies of the whole crystal when its representative point is near $A$. Similarly we can introduce normal co-ordinates at the saddle-point $B$ with associated frequencies $\nu_i'$, $i = 2, 3, \ldots 3N$. One (we have arbitrarily taken it to be the first) is missing because $V$ will decrease, rather than increase, as the associated normal co-ordinate is displaced from $B$, corresponding to unstable motion (compare $x'$ in fig. 1.3).

To compute the jump frequency for a representative point from well $A$ to well $C$ via $B$ we imagine, as in the case of the single particle, that there is a swarm (statistical ensemble) of such points. The jump frequency is then the flux of points 'from left to right' through $B$, divided by the number of points in $A$. The calculation is essentially the same as that for a particle in a two- or three-dimensional potential; the only difficulty is one of visualisation. It is fairly clear that the result will just be what one gets by extending the series (1.8), (1.15), (1.16), namely

$$f = \frac{\nu_1 \nu_2 \ldots \nu_{3N}}{\nu_2' \nu_3' \ldots \nu_{3N}'} \, \mathrm{e}^{-U/kT} \qquad (1.17)$$

where $U$ is the potential difference between $B$ and $A$.

To make contact with the macroscopic theory of diffusion we suppose that in fig. 1.3 there are $n_A$ defects in $A$, $n_C$ in $C$. The net flux of defects across $B$ from left to right is $(n_A - n_C)f$. The diffusion coefficient $D$ is the ratio of the flux of defects per unit area per unit time divided by minus the gradient of the number of defects per unit volume. If the crystal is simple cubic with lattice parameter $a$ each potential well monopolises a volume $a^3$ and the particles flow across $B$ through an area $a^2$. Hence the current is $(n_A - n_C)f/a^2$ and the defect density gradient is $(n_C - n_A)/a^4$ which gives

$$D = a^2 f. \qquad (1.18)$$

Suppose next that $n_A = n_C = n$ but that a slowly varying potential $\mathscr{V}(x)$ is added to the potential $V(x)$ of fig. 1.3. The height of the barrier from $A$ to $B$ is now $U - \frac{1}{2}a\partial\mathscr{V}/\partial x$ and the height of the barrier going from $C$ to $B$ is $U + \frac{1}{2}a\partial\mathscr{V}/\partial x$. Consequently, even though $n_A = n_C$, there is a flux from left to right over $B$ equal to

$$nf\left[\exp\left(-\frac{a}{2kT}\frac{\partial\mathscr{V}}{\partial x}\right) - \exp\left(\frac{a}{2kT}\frac{\partial\mathscr{V}}{\partial x}\right)\right] \simeq -nf\frac{a}{kT}\frac{\partial\mathscr{V}}{\partial x} \qquad (1.19)$$

provided the changes in barrier height are small compared with $kT$. If we define the drift velocity $v$ of the defects to be the flux of defects per unit area (the quantity (1.19) times $a^{-2}$) divided by the number of defects per unit volume $(n/a^3)$ we have

$$v = -\frac{D}{kT}\frac{\partial \mathscr{V}}{\partial x},$$ (1.20)

the Einstein relation between mobility and diffusion coefficient.

## 1.4 THE PHYSICAL PROPERTIES OF DEFECTS

### 1.4.1 Theory

The theory of the physical properties of point defects presents peculiar difficulties. What goes on near a defect must be found by a detailed calculation which takes into account the displacement of the ions and the redistribution of electrons. On the other hand, if one is interested in, say, the volume change due to a defect or the interaction of a defect with a dislocation some distance from it, it is impossible to take the behaviour of every atom into account and one is driven to use continuum calculations. Often the atomic and continuum approaches must be used simultaneously, suitably patched together. Atomic calculations have to begin with a careful consideration of what approximations are allowable and usually end with extensive numerical computation. They are thus hard to describe briefly and informatively, and their results are difficult to assess and liable to revision. It seems best; therefore, to confine ourselves to a few aspects which can be discussed in fair detail with simple though crude models, and only look rather cursorily at the rest.

The simplest continuum model is the so-called misfitting sphere model (Bilby, 1950). In it the point defect is treated as an elastic sphere inserted into a hole in an elastic matrix, the two being welded together at the interface (the sphere may be too large or too small). The deformation in the matrix is calculated with the help of the linear theory of elasticity, and usually with the assumption of iso-tropy. The model is obviously not a very good one, though it may have some validity when applied to alloys between metals with nearly the same atomic volume. It would be precarious to apply it to an interstitial, and in the case of a vacancy there is no misfitting atom. Even here, however, it is reasonable to suppose that the distortions round the vacancy are similar to those around a too-small misfitting atom. Despite its shortcomings we shall describe the model in some detail, since it has been, and still is, used in the theory of solid solutions and lattice defects, if only as a method of extrapolating the results of atomistic calculations to points remote from the defect, and because it underlies some of the experiments described in 1.4.2. Further it brings out some points concerning the effect of the free surface of the specimen which might be overlooked in a purely atomistic treatment.

Let $\mathbf{u}(\mathbf{r})$ be the displacement which a point of the elastic medium at $\mathbf{r}$ undergoes when the defect is introduced. According to the theory of elasticity (Love, 1952) it must satisfy the equilibrium equation

$$\mu \, \nabla^2 \mathbf{u} + (K + \tfrac{1}{3}\mu)\,\mathrm{grad}\,\mathrm{div}\,\mathbf{u} = 0 \qquad (1.21)$$

where $\mu$, $K$ are the shear modulus and bulk modulus of the (isotropic) material. The strains $e_{ij}$ are related to the Cartesian components $u_1, u_2, u_3$ of $\mathbf{u}$ by

$$e_{ij} = \tfrac{1}{2}(\partial u_i/\partial x_j + \partial u_j/\partial x_i), \quad i,j = 1,2,3$$

where $x_1$, $x_2$, $x_3$ stand for $x, y, z$. The stresses are related to the strains by

$$p_{ij} = (K - \tfrac{2}{3}\mu)\,e\delta_{ij} + 2\mu e_{ij}$$

where

$$e = e_{11} + e_{22} + e_{33} = \mathrm{div}\,\mathbf{u}$$

is the dilatation and $\delta_{ij}$ is equal to 1 or 0 according as $i$ and $j$ are or are not equal. In terms of the $p_{ij}$ the equilibrium equation (1.21) takes the form

$$\partial p_{ij}/\partial x_j = 0. \qquad (1.22)$$

(Here and in what follows we use the convention that a repeated suffix is summed over the values 1, 2, 3.)

If we require the displacement outside the defect to be spherically symmetrical and to fall off with $r$ there is only one solution of (1.21), namely

$$\mathbf{u}^\infty = c\frac{\mathbf{r}}{r^3} = -c\,\mathrm{grad}\left(\frac{1}{r}\right) \qquad (1.23)$$

where $c$ is a constant which measures the 'strength' of the defect. For many purposes it is sufficient to regard $c$ as a constant determined by experiment, or by fitting (1.23) to the results of an atomic calculation. Later we consider how $c$ is related to the details of the misfitting sphere model.

The solution (1.23) must be modified if we require it to satisfy definite boundary conditions at the surface of the solid containing the defect (see below) and so we have attached the superscript $\infty$ to **u** to indicate that it refers to an infinite medium, or more strictly to one where the inevitable boundary is so far away that we can ignore it, in considering what happens at a finite distance from the defect.

The divergence (and also the curl) of **u** is zero, so that any closed surface not embracing the defect suffers no volume change when the defect is introduced, though its shape is distorted. However, since each point of it is displaced radially outwards a surface $S$ enclosing the defect suffers an increase of volume

$$\Delta V^\infty = \int \mathbf{u} \cdot \mathbf{n} \, dS = 4\pi c \qquad (1.24)$$

(**n** is the outward normal to the surface) since the integral is just $c$ times the solid angle subtended by the closed surface at the defect. Because div $\mathbf{u}^\infty = 0$ outside the defect all such surfaces suffer the same volume change. It is sometimes useful to look at the matter in the following way. Formally the dilatation $e = \mathrm{div}\,\mathbf{u}$ is given by

$$e = -c\,\mathrm{div}\,\mathrm{grad}\left(\frac{1}{r}\right) = -c\nabla^2\left(\frac{1}{r}\right) = 4\pi c\delta(\mathbf{r}), \qquad (1.25)$$

so that $e$ vanishes except at the origin where there is a concentrated patch of dilatation of total amount $4\pi c$. An element of volume $dv$ oecomes one of volume $e\,dv$ after deformation. Consequently the volume change $\int e\,dv$ associated with any volume $V$ is $4\pi c$ or 0 according as $V$ does or does not include the defect.

Because the divergence and curl of (1.23) both vanish the associated stresses are very simple:

$$p_{ij}^\infty = 2\mu\partial u_i^\infty/\partial x_j \qquad (1.26)$$

and in particular the radial stress is

$$p_{rr}^\infty = 4\mu c/r^3. \qquad (1.27)$$

As an aside we can derive from (1.23) and (1.27) a result which will be useful later. If we cut out a sphere of radius $r$ about the defect the elastic field outside the resulting hole will be unaltered providing we maintain a pressure $p$ numerically equal to (1.27) inside it. The increase in the volume $v$ of the hole over its value when $p$ (or $c$) is zero is $\delta v = 4\pi c$. Hence

$$\frac{\delta v}{v} = \frac{p}{K_h}, \quad \text{where } K_h = \tfrac{4}{3}\mu. \tag{1.28}$$

Equation (1.28) is reminiscent of the relation

$$-\frac{\delta v}{v} = \frac{p}{K} \tag{1.29}$$

for the change of volume of a sphere under pressure, and we may think of $K_h$ as the bulk modulus for blowing up a spherical hole in an infinite solid, or a small hole in a finite solid, provided it is not too near the surface.

Returning to the main argument we must consider how these results are modified by the fact that the material has a free surface. If in the infinite medium we draw a large surface $S$ surrounding the defect and remove the material outside it, the displacement and stress will only maintain the values (1.23), (1.26) if we apply surface forces

$$T_i = p_{ij}^\infty n_j$$

per unit area of $S$, where $n_j$ is the outward normal to $S$.

If we remove these forces the displacement and stress will become, say,

$$u_i = u_i^\infty + u_i^\mathrm{I}, \quad p_{ij} = p_{ij}^\infty + p_{ij}^\mathrm{I} \tag{1.30}$$

where $u_i^\mathrm{I}$, $p_{ij}^\mathrm{I}$ is an elastic field free from singularities inside $S$ and so chosen that there are no surface tractions on $S$, that is

$$p_{ij}^\mathrm{I} n_j = -p_{ij}^\infty n_j \quad \text{on } S. \tag{1.31}$$

In analogy with electrostatics we may regard $u_i^\mathrm{I}$, $p_{ij}^\mathrm{I}$ as the 'image' elastic field. Consider first the simple case where $S$ is a sphere centred on the defect. Then according to (1.27) $p_{ij}^\infty n_j$ reduces to a uniform hydrostatic pressure $p = 16\pi\mu c/3V$ acting over the surface of the sphere, where $V$ is its volume. When this unwanted pressure is removed the sphere expands by an amount $\Delta V^\mathrm{I} = pV/K = 4\pi c(4\mu/3K)$ additional to $\Delta V^\infty$ and the whole volume expansion due to the presence of the defect is

$$\Delta V = \Delta V^\infty + \Delta V^\mathrm{I} = \gamma \Delta V^\infty = 4\pi\gamma c \tag{1.32}$$

where

$$\gamma = \frac{3K + 4\mu}{3K} = 3\frac{1 - \sigma}{1 + \sigma} \qquad (1.33)$$

($\sigma$ is Poisson's ratio).

For a body of general shape it would be difficult to work out the image field, but it is still possible to show that (1.31) remains valid. There are several ways of doing this; the following method illustrates a useful device. The total volume change is

$$\Delta V = \int_S u_j n_j \, \mathrm{d}S \qquad (1.34)$$

where $u_j = u_j^\infty + u_j^{\mathrm{I}}$. To evaluate it we replace the integrand $u_j$ by another vector $u_j'$ which leaves the integral unchanged but which has zero divergence. A suitable choice is

$$u_j' = u_j - p_{ij} x_i / 3K$$

where $u_j$, $p_{ij}$ are given by (1.30); the second term contributes nothing to the integral by (1.31) and the divergence

$$\frac{\partial u_j'}{\partial x_j} = \frac{\partial u_j}{\partial x_j} - \frac{1}{3K} p_{jj} - \frac{1}{3K} \frac{\partial p_{ij}}{\partial x_j} x_i$$

vanishes by Hooke's law and (1.22). The contribution of the unknown $u_j^{\mathrm{I}}$, $p_{ij}^{\mathrm{I}}$ to $u_i$ has zero divergence throughout the volume inside $S$ and so, by Gauss' theorem, it contributes nothing to the integral. We cannot say the same for the contribution

$$u_j^{\infty\prime} = u_j^\infty - p_{ij}^\infty x_i / 3K$$
$$= \frac{cx_j}{r^3} - \frac{2\mu}{3K} x_i \frac{\partial}{\partial x_i} \left( \frac{x_j}{r^3} \right)$$

from $u_j^\infty$, $p_{ij}^\infty$ because of the singularity at the defect. However, operating on $x_j/r^3$ with $x_i \partial/\partial x_i$ merely multiplies it by $-2$ and so $u_j^{\infty\prime} = (1 + 4\mu/3K)u_j^\infty$. Thus the integral (1.34) is $\gamma$ times the integral (1.24), which establishes (1.32) for any shape of surface and any position of the defect within it.

Since, for example, $\gamma = 1.5$ when $\sigma = \frac{1}{3}$ and $\gamma = 1.8$ when $\sigma = \frac{1}{4}$, the image correction to the expansion is quite important.

Although it is not generally possible to find the details of the change of shape induced in a body with a free surface by a single defect, it is possible to find the macroscopic change of shape when a large number of defects, say $n$ per unit volume, are scattered uniformly through it. The result is that, as is to be expected, it undergoes an increase of volume which is $n$ times the expression (1.32), but that this volume change is *uniform*, that is, though larger, the solid has the same *shape* as it did before the defects were introduced. It is not, perhaps, quite obvious that the complicated changes of shape produced by the individual defects will add up to give this simple result, and it takes a certain amount of calculation to establish it. The following is an outline of one method.

Begin by marking out the surface $S$ of the proposed solid in an infinite medium and distribute the defects uniformly within it. The displacement is now

$$u_i^\infty(\mathbf{r}) = -c \frac{\partial}{\partial x_i} \sum_m \frac{1}{|\mathbf{r} - \mathbf{r}_m|} \tag{1.35}$$

where $\mathbf{r}_m$ is the position vector of the $m$th defect. In the matrix, not too close to $S$ this may be replaced by

$$u_i^\infty(\mathbf{r}) = -cn \frac{\partial}{\partial x_i} \int \frac{dv'}{|\mathbf{r} - \mathbf{r}'|}. \tag{1.36}$$

Close to $S$ the difference between (1.35) and (1.36) will take the form of a ripple with a wave-length of order $n^{-1/3}$ which will, however, by Saint-Venant's principle, only extend into the matrix to a depth which is a few multiples of $n^{-1/3}$. We may say that (1.36) gives the macroscopic displacement in the matrix.

At this stage the volume enclosed by $S$ has been increased by $\Delta V^\infty$, for one defect, times the number of defects it contains. It has also become distorted in a way which could be calculated from (1.36) with the help of potential theory. For example, if it was originally a cube it will now be barrel-shaped ($c > 0$) or pincushion-shaped ($c < 0$). To find what happens when $S$ is freed from the matrix we appeal to a related problem.

Mark out a surface $S$ in an infinite solid, make a cut over $S$, remove the material inside and (by uniform heating or otherwise) make its volume change by a fraction $\Delta V/V$ without any change of shape or elastic constants. Force it back into the hole and weld across the cut so that points originally adjacent across $S$ are again adjacent. It is required to find the displacement in the matrix. The answer to this problem is known (Nabarro, 1940). The displacement is identical with

(1.36) provided we set

$$\Delta V / V = 4\pi \gamma c n. \tag{1.37}$$

Conversely, our distorted surface $S$ will resume its original shape when freed from the matrix but will have undergone a uniform volume expansion (1.37).

For various purposes (in particular the determination of the number of point defects from simultaneous measurements of dimensions and lattice parameter, section 1.4.2) it is necessary to know the effect of a distribution of defects on the lattice parameter as measured by X-rays.

The information obtainable about a crystal by X-ray diffraction can be summed up in a plot of scattering power in reciprocal space. For a perfect crystal this distribution takes the form of a sharp spot at each reciprocal lattice point. If the crystal is deformed (by external loads or a point defect, for example) the spots will be broadened and displaced but the new positions of their maxima will define a new deformed reciprocal lattice. Suppose that for the perfect crystal the crystal lattice points and the reciprocal lattice points are $L_i \mathbf{a}_i$ and $h_i \mathbf{b}_i$ where $\mathbf{a}_i$ and $\mathbf{b}_i$ are the base vectors of the crystal and reciprocal lattice respectively, and $L_i$, $h_i$ are integers. If the crystal is deformed the atoms move to $L_i \mathbf{a}_i + \mathbf{u} = (L_i + \Delta L_i)\mathbf{a}_i$ and the points of the reciprocal lattice to $(h_i + \Delta h_i)\mathbf{b}_i$, say, with fractional $\Delta L_i$, $\Delta h_i$. Miller and Russell (1952) have shown that for small integral $h_i$ the following relation obtains:

$$\Delta h_i A_{ij} = -h_i B_{ij} \tag{1.38}$$

where

$$A_{ij} = \sum L_i L_j, \quad B_{ij} = \sum \Delta L_i L_j$$

and the summations are over all atoms in the lattice. In what follows we shall assume for simplicity that the lattice is simple cubic with spacing $a$ and that we can replace the sums (1.38) by integrals. Then

$$A_{ij} = \frac{1}{a^5} \int_V x_i x_j \mathrm{d}v, \quad B_{ij} = \frac{1}{a^5} \int_V x_i u_j \mathrm{d}v. \tag{1.39}$$

The solution of (1.38) represents a uniform deformation of the reciprocal lattice. The dilatation of the reciprocal lattice is

$$\partial \Delta h_i / \partial h_i = \Delta h_1(100) + \Delta h_2(010) + \Delta h_3(001)$$

and its value can be inferred from measurement of reflections of not too high order. If this were all the information we had we should infer that the crystal had undergone a uniform expansion with $\Delta V/V = -\partial \Delta h_i/\partial h_i$ and we can define

$$\left(\frac{\Delta V}{V}\right)_{\mathrm{X}} = -\frac{\partial \Delta h_i}{\partial h_i} \qquad (1.40)$$

to be the volume change of the crystal as measured by X-rays.

It is obvious from the form of $B_{ij}$ that $(\Delta V/V)_{\mathrm{X}}$ depends on a quite different average of the displacements from $\Delta V/V$ which depends on the surface integral (1.34). Consequently if $u_i$ is due, say, to a single defect in the crystal we do not expect $(\Delta V/V)_{\mathrm{X}}$ and $\Delta V/V$ to agree. It turns out, however, that, as might be expected, if there are a large number of defects distributed uniformly through the crystal they do agree, and that, moreover, the detailed deformation of the reciprocal lattice near its origin is what one would have inferred from its macroscopic change of shape.

These points may be shown quite easily for a sphere containing a single defect at $\boldsymbol{\xi}$, so that the displacement is given by (1.23) with $\mathbf{r}$ replaced by $\mathbf{r} - \boldsymbol{\xi}$, together with the image field necessary to give zero surface tractions.

For a sphere of radius $R$, $A_{ij} = (4\pi R^5/15a^5)\delta_{ij}$ and so

$$\left(\frac{\Delta V}{V}\right)_{\mathrm{X}} = -\frac{\partial \Delta h_i}{\partial h_i} = \frac{15}{4\pi R^5} \int x_i u_i \, \mathrm{d}v.$$

But

$$\int x_i u_i \, \mathrm{d}v = \int_V x_i u_k \frac{\partial x_i}{\partial x_k} \, \mathrm{d}v = \int \left[ \frac{\partial}{\partial x_k} (r^2 u_k) - er^2 - x_i u_i \right] \mathrm{d}v$$

so that

$$2\int_V x_i u_i \, \mathrm{d}v = \int_S r^2 u_k n_k \, \mathrm{d}S - \int_V er^2 \, \mathrm{d}v.$$

The first integral is just $R^2 \Delta V = R^2(\Delta V^\infty + \Delta V^{\mathrm{I}})$. In the second, $e^\infty$ contributes $4\pi c\xi^2$ by (1.25), where $\xi$ is the distance of the defect from the centre of the sphere. To find the contribution of $e^{\mathrm{I}}$ to the second integral we note that if the defect moves over the sphere $\xi = \mathrm{const.}$ the image field will move round with it but the volume integral of $e^{\mathrm{I}}r^2$ will be unaltered. Hence the defect may be replaced by a uniform distribution of infinitesimal defects over the sphere $\xi = \text{const.}$, provided their total strength is $c$. But the image dilatation is then, by symmetry, uniform and equal to $\Delta V^{\mathrm{I}}/V$ and this gives an image contribution of $\frac{3}{5}R^2 \Delta V^{\mathrm{I}}$ to the volume integral. Hence finally, using (1.32),

$$\left(\frac{\Delta V}{V}\right)_{\mathrm{x}} = \left[1 + \frac{1}{2\gamma}\left(3 - 5\frac{\xi^2}{R^2}\right)\right]\frac{\Delta V}{V} \qquad (1.41)$$

where $\Delta V$ is the geometrical volume change, equation (1.32). For a defect at the centre of the sphere the ratio $(\Delta V/V)_{\mathrm{x}}/(\Delta V/V)$ is $1 + 3/2\gamma \sim 2$, and for a defect near the surface it is $1 - 1/\gamma \sim 0.3$. (If we had ignored the image terms the discrepancy would have been greater, with a ratio 2.5 for a defect at the centre and 0 for one near the surface, for any Poisson's ratio.) Consequently, if defects are not distributed uniformly through the crystal the volume changes inferred from X-rays and changes in dimensions or density will not agree. However, if they are distributed uniformly $(\Delta V/V)_{\mathrm{x}}$ is found from (1.41) by multiplying by $n$, the number of them in unit volume, and replacing $\xi^2$ by its mean value over the sphere, $\frac{3}{5}R^2$. The square bracket expression then becomes unity, so that the X-ray and geometrical volume expansions are equal. Also, by symmetry the reciprocal lattice will undergo a *uniform* contraction at least near the origin.

By more elaborate calculations it may be shown that for an arbitrary shape of crystal with a uniform distribution of defects $(\Delta V/V)_{\mathrm{x}} = \Delta V/V$ and that the reciprocal lattice is deformed in the way which the macroscopic deformation of the crystal would suggest.

So far, we have not needed to know the relation between the strength $c$ of a defect and the details of the misfitting sphere model but we shall require it in section 1.5. The connection is easily made.

If $V_{\mathrm{s}}$, $V_{\mathrm{h}}$ are the volumes of sphere and hole before insertion, we define

$$V_{\mathrm{mis}} = V_{\mathrm{s}} - V_{\mathrm{h}}$$

to be the misfit volume. When the sphere is inserted in the hole $V_{\mathrm{s}}$ and $V_{\mathrm{h}}$ will change by $\Delta V_{\mathrm{s}}$ and $\Delta V_{\mathrm{h}}$ and pressures will be developed on the surfaces of the hole and sphere. Since these must be equal, (1.28) and (1.29) give

$$\tfrac{4}{3}\mu\Delta V_{\mathrm{h}} + K'\,\Delta V_{\mathrm{s}} = 0 \qquad (1.42)$$

where $K'$ is the bulk modulus of the sphere. (We ignore the slight difference between the initial volumes of hole and sphere.) Also, $\Delta V_h$ and $\Delta V_s$ must be just such as to eliminate $V_{mis}$, that is,

$$\Delta V_h - \Delta V_s = V_{mis}. \qquad (1.43)$$

Further, $\Delta V_h$, being the volume change of a sphere closely enveloping the defect, is equal to $\Delta V^\infty$. Equations (1.42) and (1.43) give

$$\Delta V_h = \Delta V^\infty = 4\pi c = V_{mis}/\gamma'$$

where

$$\gamma' = \frac{3K' + 4\mu}{3K'}.$$

From the previous argument, the total volume change is

$$\Delta V = \gamma V_{mis}/\gamma'. \qquad (1.44)$$

If the sphere has the same bulk modulus as the matrix $\Delta V = V_{mis}$, while if it is rigid ($K' = \infty$) $\Delta V = \gamma V_{mis}$.

For a vacancy there is nothing in the hole to cause a volume change. In this case we can follow Brooks (1955) and regard the vacancy as a hole with a radius $r_s$, defined by $\frac{4}{3}\pi r_s^3 = \Omega$, pulled inwards by the macroscopic surface tension $\Gamma$. The volume change can be found from (1.28) with $p = -2\Gamma/r_s$. This gives $\Delta V^\infty$, so that the total volume change is given by

$$\Delta V/\Omega = -3\gamma\Gamma/2\mu r_s.$$

The ratio $\Gamma/\mu$ has the dimensions of a length. For a wide range of materials it has a value close to $\frac{1}{4}$Å (Nabarro, 1951). For copper $r_s = 1.4$ Å, giving $\Delta V/\Omega = -0.4$ which is in quite good agreement with the values $-0.53$ (Tewordt, 1958), $-0.29$ (Seeger and Mann, 1960) and $-0.48$ (Johnson and Brown, 1962) obtained by atomistic calculations. However, the same model over-estimates the energy of formation by a factor of about two so it cannot be taken very seriously. In fact, the energy of formation is $4\pi\Gamma r_s^2$ plus the elastic energy in the matrix. The latter is $\frac{1}{2}p\delta v$ calculated from (1.28) with $p = -2\Gamma/r_s$. Its ratio to the surface term is $\Gamma/2\mu r_s$. With $\Gamma/\mu \sim \frac{1}{4}$ Å and $r_s = 1.5$ Å the ratio is about 0.1, so that we may take the energy to be simply $4\pi\Gamma r_s^2$. For copper ($\Gamma = 1400$ dynes cm$^{-1}$) this gives about 2 eV, which is about twice the value determined from experiment and more respectable calculations.

This crude model actually contains an element of truth. If we cut a large hole in a piece of metal the electrons will tunnel a certain distance into the cavity to an extent governed by the fact that though they thereby reduce their kinetic energy they also increase their potential energy. These ideas can be made the basis of a calculation of surface energy (Huang and Wyllie, 1949). The same principles apply when the hole is reduced to one atomic volume (vacancy). The tunnelling length is then comparable with the dimensions of the cavity; as a result, the increase of energy is less than the macroscopic model would suggest. Detailed calculations on these lines have been made by Fumi (1955) and Seeger and Bross (1956). As the surface tension model suggests, the 'elastic' part of the energy (ion–ion interaction) is not very important.

The electrical resistivity due to point defects can be calculated from the formula (Mott and Jones, 1936)

$$\Delta\rho = \frac{mv}{ne^2}cA$$

where $m$ is the electron mass, $v$ the Fermi velocity, $n$ the number of free electrons per atom, $e$ the electronic charge and $c$ the atom fraction of defects. The quantity $A$ is the scattering cross-section given by

$$A = \int (1 - \cos\theta)f(\theta)\,d\omega$$

where $f(\theta)$ is the intensity of the scattered wave in the direction $\theta$. If the scattered wave is analysed into a set of products of spherical harmonics of order $l$, each multiplied by a radial factor, $A$ can be written in terms of the phase shifts $\eta_l$ of the various components (Mott and Massey, 1965):

$$A = \frac{4\pi}{k^2}\sum_l (l+1)\sin^2(\eta_l - \eta_{l+1})$$

where $k = 2mv/\hbar$.

The phase shifts must satisfy the Friedel (1952) sum rule which requires that the relation

$$\frac{2}{\pi}\sum_l (2l+1)\eta_l = Z$$

holds, where $Z$ is the charge required to screen the potential of the defect. This is a powerful check on the results obtained from an assumed potential. Indeed Abelès (1953) was able to obtain a reasonable value for vacancy resistivity simply by taking for the potential a flat-bottomed spherical well of one atomic volume and fixing its depth so that the sum rule was satisfied. In most other calculations a screened Coulomb potential has been used.

The results of a number of calculations suggest that the extra resistance due to one per cent of vacancies in copper, silver or gold is about 1.5 microhm cm, and that the contribution from lattice distortion is small.

### 1.4.2 Experimental measurement

Individual point defects can be seen by field-ion microscopy (Müller, 1962; Brandon and Wald, 1961). This technique has limitations (in particular, the defects are only observed at the surface) and usually one must study defects indirectly through their effect on some bulk property. An obvious way to do this is to vary the temperature, so that the defect concentration varies according to (1.3), and then measure the change in some bulk property supposedly proportional to the defect concentration. It is, of course, necessary to allow for the changes in the chosen property which would occur even if there were no defects. In the present section we show how this difficulty is overcome in a few typical experiments. First, however, it is necessary to say something about the relation between the free energy, enthalpy and entropy of formation.

Suppose that the concentration of some type of defect is given by

$$c = \mathrm{e}^{-\Delta G/kT} \tag{1.45}$$

and that there is some physical effect proportional to the concentration of defects, so that $X = Ac$ where $A$ is independent of temperature or can be corrected for temperature variation. If the results of a series of measurements of $X$ as a function of temperature are exhibited by plotting $-\ln X$ against $1/kT$, the slope of the resulting curve is not $\Delta G$ but rather the quantity $\Delta H$ related to it by

$$\Delta H = \frac{\partial}{\partial(1/T)} \cdot \frac{\Delta G}{T} = \Delta G - T \frac{\partial \Delta G}{\partial T} . \tag{1.46}$$

Equation (1.46) is identical with the Gibbs–Helmholtz relation connecting Gibbs free energy and enthalpy, so that $\Delta H$ is the enthalpy of formation, or heat of formation. If the material is not under pressure it is just the (internal) energy of formation. Again, according to thermodynamics the negative temperature derivative of a Gibbs free energy is the associated entropy, so that

$$\Delta S = -\frac{\partial \Delta G}{\partial T}$$

is the entropy of formation, and we may write

$$\Delta G = \Delta H - T\Delta S. \tag{1.47}$$

The entropy of formation can be related to the change which the vibration frequencies of the lattice undergo when it contains a defect. Suppose that in fig. 1.3 (regarded, as before, as a schematic representation of the configuration space describing the whole crystal) $A$ represents the perfect crystal and $C$ represents the state of metastable equilibrium in which there is, say, a vacancy at some specific lattice site. The representative points can pass from $A$ to $C$ or from $C$ to $A$ by way of the saddle-point $B$. If $n_A$ and $n_C$ are the numbers of representative points in $A$ and $C$ and $f_A$, $f_C$ are the corresponding jump frequencies we must have $n_A f_A = n_C f_C$ in equilibrium.

The jump frequency $f_A$ is given by (1.17) with $\nu_i = \nu_i^A$ and $U = V_B - V_A$, and $f_C$ by the same expression with $\nu_i = \nu_i^C$ and $U = V_B - V_C$. When we equate $n_A f_A$ with $n_C f_C$ the saddle-point frequencies $\nu_i$ and potential $V_B$ cancel out and we are left with

$$n_C/n_A = (\Pi\nu_i^A/\Pi\nu_i^C)\,e^{-V/kT} \tag{1.48}$$

where $V$ is the difference in potential between $C$ and $A$, and the products extend over the $3N$ normal vibration frequencies of the perfect and defective crystal. But the ratio $n_C/n_A$ represents the relative probability of finding or not finding a vacancy at the specified site. Comparing (1.48) with (1.45) and (1.47) we see that we must put $V = \Delta H$ and

$$\Delta S = k\ln(\Pi\nu_i^A/\Pi\nu_i^C).$$

The argument, of course, also applies to defects other than vacancies.

If, as we have supposed, the crystal executes purely harmonic vibrations about its perfect and defective states $\Delta H$, $\Delta S$ are independent of temperature. (The fact that we have to go through a highly non-linear region to get from one state to the other is irrelevant.) Actually, there are anharmonic and other effects which spoil this, but they are small and tend to cancel (Levinson and Nabarro, 1967) and $\Delta H$, $\Delta S$ are commonly treated as constants. We can then write

$$c = e^{\Delta S/k}\, e^{-\Delta H/kT}$$

where $\exp(\Delta S/k)$ is the so-called entropy factor. Then if the proportionality factor $A$ connecting the physical effect $X$ with the concentration $c$ is known both $\Delta H$ and $\Delta S$ can be found from a logarithmic plot, but if $A$ is not known only $\Delta H$ can be determined.

In practice, the variation of $X$ due to the change of defect concentration with temperature will be obscured by the variation of $X$ which would occur if defect formation could somehow be inhibited. For example, the volume change due to defects would be swamped by thermal expansion, and the defect resistivity by the normal change of resistance with temperature.

There are three ways round this difficulty: (i) we may estimate the non-defect part of $X$ by extrapolation from temperatures where the defect contribution can reasonably be neglected; (ii) the non-defect contribution can be eliminated by some special artifice; (iii) we may quench specimens from a number of temperatures down to a single common temperature. If the quench has been fast enough to retain all the defects and if additional defects have not been introduced as a result of quenching stresses we then have a set of specimens containing the defect concentration appropriate to the temperatures from which they were quenched. A plot of $X$ measured at the low temperature against the quenching temperature will then give information about the variation of vacancy concentration with temperature.

An example of the first method is provided by MacDonald's (1954) measurement of the defect contribution to the specific heat of the alkali metals. The presence of defects of concentration $c$ and heat of formation $\Delta H$ will give an extra contribution to the specific heat per atom equal to $\partial(\Delta Hc)/\partial T$. According to (1.45) and (1.46) this amounts to

$$\Delta C_p = \frac{(\Delta H)^2}{kT^2} c$$

if we suppose that $\Delta H$ is temperature-independent. Fig. (1.4) shows the experimental curve for potassium; it gives $\Delta H = 0.4$ eV.

The second method is illustrated by Simmons and Baluffi's (1960a, b, 1962, 1963) experiments in which the change in both length and X-ray lattice parameter of a specimen were measured. To explain their method we suppose for the moment that there is no relaxation

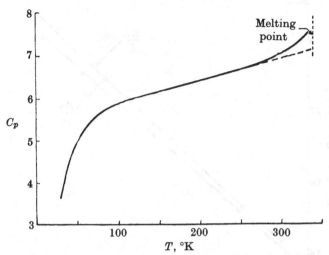

Fig. 1.4. The specific heat of potassium as a function of temperature. (MacDonald, 1954)

of the lattice about a vacancy or interstitial. Then each vacancy increases the volume of the specimen by one atomic volume, and each interstitial decreases it by the same amount, and so a fractional concentration $c_v$ of vacancies and $c_i$ of interstitials produces a fractional volume change $c_v - c_i$. However, if we vary the temperature the volume change due to $c_v - c_i$ will be overshadowed by thermal expansion. On the other hand, if there is no relaxation the X-ray lattice parameter is only affected by thermal expansion and not by the presence of the defects. This is just an extreme case of the fact (Huang, 1947) that altering the scattering power of some of the atoms in a lattice does not affect the X-ray parameter; the scattering power of some lattice points is reduced to zero (vacancies) and the scattering power of some interstitial sites is increased from zero to a finite value (interstitials). Hence if there were no relaxation we should have

$$c_v - c_i = \frac{\Delta V}{V} - \left(\frac{\Delta V}{V}\right)_x = \frac{\Delta V}{V} - 3\frac{\Delta a}{a} \qquad (1.49)$$

where $\Delta a/a$ is the fractional change in lattice parameter. But we have seen in section 1.4 that if relaxation does occur it affects $\Delta V/V$ and $(\Delta V/V)_x$ equally, so that (1.49) is still valid.

Figure 1.5 shows some of Simmons and Baluffi's experimental results for gold. Even a rough estimate suggests that $\Delta H$ is much greater

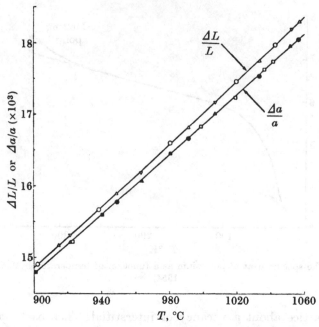

Fig. 1.5. Change of length ($L$) and lattice parameter ($a$) of copper with temperature. (Simmons and Baluffi, 1962)

for interstitials than it is for vacancies, so we may expect $c_i$ to be negligible in comparison with $c_v$. Consequently a plot of the logarithm of $\Delta V/V - 3\Delta a/a$ against $1/kT$ will give both the energy and entropy of formation of vacancies. The following are some values obtained in this way by Simmons and Baluffi:

|              | Cu    | Ag    | Au   | Al   |
|--------------|-------|-------|------|------|
| $\Delta H$ (eV) | 1.17  | 1.09  | 0.94 | 0.75 |
| $\Delta S/k$    | ~1.5  | ~1.5  | 1    | 2.4  |

The estimated error in $\Delta H$ is about 10 per cent.

The quenching experiments of Bauerle and Koehler (1957) are a classic example of the third method. Fig. 1.6 shows their results for the quenched-in resistance of gold wires. The energy of formation (0.98 ± 0.03 eV) agrees closely with the value which Simmons and Baluffi obtained from equilibrium measurements. Since Simmons and Baluffi's measurements give the absolute concentration of vacancies they can be combined with Bauerle and Koehler's results to give the resistance increase per vacancy. The result is 1.8 microhm cm. for

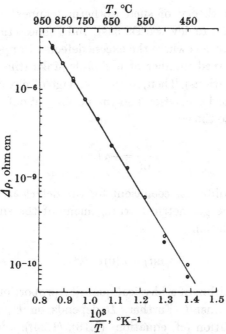

Fig. 1.6. Variation of quenched-in resistivity ($\rho$) of copper with quenching temperature. (Bauerle and Koehler, 1957)

one per cent vacancies, which agrees reasonably with theoretical estimates.

Huebner and Homan (1963) have found the volume increase associated with the formation of a vacancy in gold. If we apply the thermodynamic relation $\Delta V = \partial\Delta G/\partial p$ to (1.45) we get

$$\frac{\partial \ln c}{\partial p} = -\frac{\Delta V}{kT}.$$

Huebner and Homan subjected gold wires to a temperature of 680 °C and pressures up to 11 000 atmospheres, quenched them and measured the electrical resistance. The slope of the resistance–pressure curve gave for $\Delta V$ a fraction $0.53 \pm 0.04$ of an atomic volume, so that, of the basic increase of one atomic volume per defect, about half is cancelled by relaxation.

If a specimen containing quenched-in vacancies or other point defects is held at a temperature between the temperatures from and to which it was quenched the excess vacancies will gradually disappear until the concentration appropriate to the new temperature is reached. The theory of such isothermal annealing processes is complicated; for a review see Damask and Dienes (1963).

The simplest case is where the excess defects disappear by random migration to a fixed number of unfillable sinks (the surface of the specimen, dislocations). Then, in the language of physical chemistry, we have a first-order reaction and the number of defects varies with time according to the law

$$\frac{dn}{dt} = -\alpha D n \qquad (1.50)$$

where $D$ is the diffusion coefficient for the defect and $\alpha$ is a factor depending on the geometrical arrangement of the sinks. Equation (1.50) has the solution

$$n(t) = n(0)\,e^{-\alpha D t}$$

so that the logarithm of a physical quantity proportional to $n$ should plot linearly against $t$. Further, $D$ depends on $U$, the activation energy for migration (cf. equations (1.18), (1.20)), and so $U$ can be determined from the results of isothermal annealing experiments at several temperatures. (It is not necessary to know the parameter $\alpha$.)

In this way Bauerle and Koehler (1957) found the migration energy for vacancies in gold by measuring the decay of the extra resistance. They obtained $0.82 \pm 0.05$ eV. This can be checked by comparison with measurements of the coefficient of self-diffusion. Since self-diffusion occurs by the interchange of vacancies it is proportional to the product of the concentration of vacancies and the diffusion coefficient for vacancies and so contains a factor $\exp[-(U + \Delta H)/kT]$. Diffusion experiments (Makin et al., 1957) give $U + \Delta H = 1.8$ eV which agrees closely with the value 1.80 eV found by combining the $U = 0.82$ eV of Bauerle and Koehler's annealing experiments with the $\Delta H = 0.98$ eV obtained from their quench experiments.

Stored energy measurements can give an absolute value for vacancy concentration. De Sorbo (1960) quenched a series of gold specimens from different temperatures and measured calorimetrically the release of energy as the excess defects (assumed to be vacancies) annealed out. A logarithmic plot gives the energy of formation, and this, since it is the energy released per defect annihilated, can be used to convert the energy measurements into vacancy concentrations.

His result was

$$c = 2\exp(-0.97\,\text{eV}/kT).\qquad(1.51)$$

The energy of formation agrees well with the values obtained from equilibrium measurements, though the pre-exponential factor is rather smaller. In conjunction with the resistivity annealing experiments just described Bauerle and Koehler also measured length changes. Combined with (1.51) their results give 0.57 atom volume for the volume increase due to a vacancy, in fair agreement with the value Huebner and Homan obtained from their pressure-dependence measurements.

All the experiments described above refer to thermally generated defects, either in equilibrium or retained by quenching. Similar methods can also be used to study defects introduced by irradiation or plastic deformation. We give a couple of examples.

If irradiation produces equal numbers of vacancies and interstitials we expect, according to (1.49) that the fractional changes in length and X-ray lattice parameter will be equal. This has been verified for copper irradiated with deuterons at low temperatures (Simmons and Baluffi, 1958; Vook and Wert, 1958).

Molenaar and Aarts (1950) measured the rate of change of electrical resistance with deformation in copper at liquid nitrogen temperature. They estimated the number of defects by measuring the resistance decrease on annealing at room temperature. If the defects are assumed to be vacancies producing a resistivity change of 1.5 microhm cm. per atom per cent vacancies, their results give a defect concentration of $4 \times 10^{-6}$ for one per cent deformation. Henderson and Koehler (1956) measured the stored energy released on annealing copper deformed at a low temperature. With an energy of formation of 1 eV per defect their results give a concentration of $6 \times 10^{-6}$ defects for 1 per cent deformation. Both these values are comparable with the estimate at the end of section 1.2.

## 1.5 INTERACTION ENERGIES OF POINT DEFECTS

In addition to producing the macroscopic effects which enable their fundamental parameters to be measured (section 1.4) point defects play an important role in influencing the mechanical properties of metals, chiefly through their interaction with dislocations. In this section we discuss the basic interactions between point defects and the stresses due to dislocations, other defects, or external loading.

Consider a point defect in a crystal in a state of stress, due to other defects, dislocations and externally applied forces. If the point defect is moved about, the total energy of the system, that is, the sum of the internal energy (strictly the Helmholtz free energy) and the potential of the external loading mechanism, varies. We may regard the negative gradient of the total energy as a 'force' acting on the defect. From this force one can calculate at what points the defect will be in equilibrium, and with the help of the Einstein relation (1.20) the drift velocity of a defect can be calculated.

If the problem is treated according to the infinitesimal theory of elasticity it is usually convenient to consider not the total energy but only the cross term in it between the elastic field of the defect and the applied internal stresses which are regarded as exerting a force on it. In this section we treat directly some cases which are important in the theory of the interaction between point defects and dislocations. It is possible to develop a fairly respectable general theory of the forces on lattice defects (see, for example, Eshelby, 1956) which shows that some of the shortcomings of the following treatment are only apparent. For example, equation (1.52) below states that the force on a point defect in the elastic field of a dislocation depends on the value of the hydrostatic pressure due to the latter at the position of the defect, where things are highly non-superposable. The precise treatment shows that for the result to be valid it is only necessary that one can find a closed surface, isolating the defect from the dislocation, on which the elastic fields of dislocation and defect can be superposed with fair accuracy. The appropriate hydrostatic pressure to use is then that found by extrapolating inwards from this surface to the position of the defect, but on the assumption that the defect is absent. Likewise, interaction energies derived on the assumption that the applied stress is uniform remain valid when it is not, provided the applied field near the defect (in the sense just described) does not vary too rapidly. We begin with the simple case of a point defect in a crystal under hydrostatic pressure.

We may define the interaction energy of a defect in a specimen subjected to a hydrostatic pressure $p$ to be the work required to insert the defect in the specimen under pressure $p$, minus the work required to insert it when $p = 0$. The latter is $E_s$, the self-energy of the defect. The former is $E_s + p\varDelta V$, the second term being simply the work done against $p$ by the increase $\Delta V$ (equation (1.32)) in the volume of the specimen. Hence

$$E_{\text{int}} = p\Delta V. \tag{1.52}$$

One might expect that there would be an additional term arising from a cross-term in the elastic energy of the specimen between the elastic field of the defect and the applied stress, but in fact this term is zero. The elastic energy is

$$\tfrac{1}{2}\int (p_{ij}^{\text{D}} + p_{ij}^{\text{A}})(e_{ij}^{\text{D}} + e_{ij}^{\text{A}})\,\mathrm{d}v \tag{1.53}$$

taken over the volume of the specimen. The superscripts A and D refer to the applied field and the defect. The A-A term represents the energy when $p$ is applied in the absence of the defect and the D-D term is the self-energy of the defect. The cross-term is

$$\tfrac{1}{2}\int (p_{ij}^{\text{D}} e_{ij}^{\text{A}} + p_{ij}^{\text{A}} e_{ij}^{\text{D}})\,\mathrm{d}v. \tag{1.54}$$

If the two terms are written out in terms of displacements they are seen to be identical, so that (1.54) may be replaced by twice its first term. From the definition of the strain components, and because $p_{ij} = p_{ji}$, we may write

$$p_{ij}^{\text{D}} e_{ij}^{\text{A}} = \tfrac{1}{2}p_{ij}^{\text{D}}(\partial u_i^{\text{A}}/\partial x_j + \partial u_j^{\text{A}}/\partial x_i) = p_{ij}^{\text{D}}\,\partial u_i^{\text{A}}/\partial x_j$$

and since $\partial p_{ij}^{\text{D}}/\partial x_i = 0$ we may also write it as $\partial(p_{ij}^{\text{D}} u_i^{\text{A}})/\partial x_j$. In this form the volume integral can be converted by Gauss' theorem into a surface integral

$$\int p_{ij}^{\text{D}} u_i^{\text{A}} n_j \,\mathrm{d}S \tag{1.55}$$

extended over the surface of the specimen. But $p_{ij}^{\text{D}} n_j$ vanishes at the surface so that the cross-term is zero. Since the displacement is discontinuous across the interface between defect and matrix we should, perhaps, for safety's sake, also extend the integral (1.55) over surfaces just outside and just inside the interface. However, since $p_{ij}^{\text{D}} n_j$ and $u_i^{\text{A}}$ (though not $u_i^{\text{D}}$) are continuous across the interface and the two surfaces have oppositely-directed normals they contribute nothing.

If the defect is a misfitting sphere with the same elastic constants as the matrix this concludes the proof that (1.52) is the appropriate interaction energy. With a suitable interpretation it is also correct when the defect is both misfitting and inhomogeneous, with elastic constants differing from those of the matrix. Suppose that at first the sphere is inhomogeneous but perfectly fitting. When the pressure is applied the work done in compressing the solid will not be the same as it would be in the absence of the defect because the presence of the defect makes the specimen harder or softer. There is thus an interaction energy between an applied stress and a perfectly fitting but inhomogeneous defect. It is discussed below and shown to be proportional to $p^2$. Suppose now that the defect is made to become misfitting. The work required to do this is the sum of $E_\mathrm{S}$, the expression (1.52) and the cross-term in (1.53). In the latter, $p_{ij}^\mathrm{A}$, $e_{ij}^\mathrm{A}$ now refer to the elastic field set up by the applied forces in the material containing the inhomogeneous but perfectly fitting defect. The cross-term vanishes by the previous argument, which nowhere assumes the material to be homogeneous. Consequently (1.52) gives the interaction energy of the defect in its role of a misfitting inclusion (source of internal stress) over and above its interaction energy as an inhomogeneity. Since the latter quantity is quadratic in the applied stresses, the linear interaction term is always given correctly by (1.52).

Equation (1.52) can be re-written in terms of the variation of lattice parameters $a$ with defect concentration $c$. Since $\Delta V = (3/a)$ $(\mathrm{d}a/\mathrm{d}c)\Omega$ where $\Omega$ is the atomic volume, (1.52) becomes

$$E_\mathrm{int} = \frac{3}{a}\frac{\mathrm{d}a}{\mathrm{d}c}\,\Omega p. \qquad (1.56)$$

This remains true even if we drop the limitation to spherically symmetrical deformation about a defect and isotropic elasticity, provided a uniform distribution of defects produces a volume change in the crystal but no macroscopic change of shape.

A uniform distribution of defects like the one in fig. 1.1(b) (or an even less symmetrical one), all with the same orientation, will change both the size and shape of a crystal. The change can be specified by the strain $e_{ij}^\mathrm{T}(c)$ which a unit cube suffers when a concentration $c$ is introduced. If the cube is subjected to an applied stress $p_{ij}$ the work done against it when the defects are introduced is $-e_{ij}^\mathrm{T}p_{ij}$ and so instead of (1.56) we have for the interaction energy per defect

$$E_\mathrm{int} = -\frac{\mathrm{d}e_{ij}^\mathrm{T}}{\mathrm{d}c}\,p_{ij}\Omega; \qquad (1.57)$$

the cross-term in the internal energy vanishes by the same argument as before. The difference in sign between (1.56) and (1.57) is merely a result of the convention that for a hydrostatic pressure $p$ the stress components are $p_{11} = p_{22} = p_{33} = -p$, $p_{12} = p_{23} = p_{31} = 0$. If we insert these values in (1.57) we get back to (1.56) because $e_{11}^T + e_{22}^T + e_{33}^T$ is just the fractional change of volume. Conversely, if the defects have spherical or cubic symmetry $e_{11}^T = e_{22}^T = e_{33}^T$, $e_{12}^T = e_{23}^T = e_{31}^T = 0$ and (1.57) reduces to (1.56) for any state of stress; the defect only pays attention to the hydrostatic component of the $p_{ij}$. The validity of (1.57) does not depend on the material being isotropic.

An important application of (1.52) is to the interaction between a defect and an edge dislocation. Here the hydrostatic pressure is

$$p = \frac{\mu b}{\pi \gamma} \frac{\sin \theta}{r} \qquad (1.58)$$

where $r$, $\theta$ are polar co-ordinates with $\theta$ measured from the direction of the Burgers vector $b$, $\gamma$ is given by (1.33) and the material is assumed to be isotropic. Equation (1.52) gives

$$E_{\text{int}} = \frac{\mu b}{\pi \gamma} \Delta V \frac{\sin \theta}{r} = \frac{\mu b}{\pi} \Delta V^{\infty} \frac{\sin \theta}{r}. \qquad (1.59)$$

Cottrell (1948) first calculated the interaction between a point defect and an edge dislocation by considering the work done in 'blowing up' the defect in the field of the dislocation. He obtained the expression (1.59) multiplied by $\gamma$; Bilby (1950) later obtained the correct result. Since Cottrell considered a rigid sphere while Bilby considered one elastically homogeneous with the matrix it is commonly stated that the difference arises from the strain energy inside the sphere. However, we have seen that the linear part of the interaction energy should be the same, for the same $\Delta V$ or $\Delta V^{\infty}$, whether the sphere is rigid (when there is no strain energy within it) or not. One can also be confused by the fact that the ratio of Cottrell's result to that of Bilby, the ratio of $\Delta V$ to $\Delta V^{\infty}$ for a particular defect, and the ratio of the $\Delta V$s (or $\Delta V^{\infty}$s) for two defects with the same volume misfit, the one rigid, the other elastically homogeneous (cf. equation (1.44)), all differ by a factor $\gamma$, and that the expression (1.58) itself contains a factor $\gamma^{-1}$.

Cottrell's method has the advantage that it is only concerned with what goes on near the defect and so avoids the need to take the not very obvious step of applying (1.52), derived for an externally applied stress, to the case of internal stress. For the former, it is fairly obvious why the total change of volume $\Delta V$ appears in (1.52) rather than $\Delta V^{\infty}$, but it is not at all obvious why it should also appear in the interaction between a defect and a dislocation unless we appeal to the anthropomorphic principle that the defect does not know what is producing the pressure. It is therefore perhaps worthwhile to show that, with attention to a few details, Cottrell's method can be made to give the correct result. We limit ourselves to the case of an incompressible sphere. There is nothing specific to the dislocation problem in the following treatment: it applies equally well to any source of internal stress or external loading.

We note first that to correspond with a hole of volume $V_h$ in the unstrained matrix we must cut out a hole of volume

$$V_h' = V_h(1 - p/K)$$

in the medium under pressure $p$, and then put a pressure $p$ in the hole to prevent it collapsing further. Next, as the hole is inflated from the initial volume $V_h'$ to a volume $V_s$ which will enable it to accept the rigid sphere, an additional pressure $(4\mu/3V_h)(V_s - V_h')$ will build up, equal to the product of the fractional change in volume and the bulk modulus for blowing up a hole (equation (1.28)). The final pressure is thus

$$p_f = \frac{4\mu}{3} \cdot \frac{V_s - V_h'}{V_h} + p$$

while the initial pressure is $p$. The work $W$ done in inserting the sphere is the product of the average pressure and the change of volume:

$$W = \tfrac{1}{2}(p_f + p)(V_s - V_h'). \tag{1.60}$$

When this is worked out the term linear in $p$ is found to be $p\gamma(V_s - V_h)$ which agrees with (1.52) since $V_s - V_h = \Delta V^{\infty}$ for a rigid sphere.

We can also use (1.60) to find the work required to insert the rigid sphere correct to order $p^2$. For consistency, however, we must first subtract from it the energy recovered on relaxing the stress in the matrix atom which is removed to make room for the rigid sphere, namely $\tfrac{1}{2}p(V_h - V_h')$. The result is

$$E = \tfrac{2}{3}\mu \frac{(\Delta V^{\infty})^2}{V_h} + \gamma \Delta V^{\infty} p + \tfrac{1}{2}\frac{\gamma V_h}{K} p^2.$$

The first term, independent of $p$, is the self-energy of the defect. The second we have already discussed. The third is independent of the degree of misfit and appears because the sphere is an inhomogeneity in the elastic constants. Evidently (at least if $\Delta V^{\infty} = 0$) it will be energetically advantageous for the defect to migrate to a region where $p^2$ is smaller. There will obviously be a similar though smaller effect if the sphere, though not absolutely rigid, has a greater bulk modulus than the matrix. Conversely, we should expect that vacancies or elastically softer atoms would tend to migrate to positions of high $p^2$. Also, we should expect similar effects in a more general stress field, including a pure shear stress.

All these effects can be calculated on the inhomogeneous sphere model but, as in the case of (1.56) and (1.57), it is more satisfactory to relate them directly to an associated macroscopic effect, in this case the change of the bulk elastic constants which the presence of defects induces in the solid.

Consider a rod containing a uniform distribution of point defects which, say, reduce its elastic constants. Let one end of it be clamped in a rigid vice so as to impose a fixed strain (not stress). The energy density at the clamped end is proportional to the square of the imposed strain times a suitable elastic constant. So if the impurities migrate to the clamped end and reduce the elastic constants there, the elastic energy will be decreased. Consequently we can say that there is a force urging the individual defects towards the clamped end. Suppose that instead of clamping we impose invariable surface forces (dead loading) chosen so as to produce a region of fixed stress (not strain) at one end of the rod. Then the energy density, being proportional to the square of the imposed stress divided by a suitable elastic constant, will be *increased* if the defects migrate to the stressed end. Nevertheless, it will still be energetically favourable for them to do so because the stressed end becomes elastically softer and this allows the applied forces to do more work on the bar (Le Chatelier's principle). If the stress–strain relation is strictly linear the work done by the applied forces is precisely twice the increase in elastic energy and there is a net reduction of energy.

To make these ideas quantitative suppose that we can at will force a unit cube of material into a rigid container which is nearly a cube, but with angles differing slightly from right angles (so as to induce a shear strain) and which has a volume slightly less than unity (so as to induce a hydrostatic pressure). Insert the defect-free cube of material. The elastic energy is $W_d + W_s$ where $W_d$ and $W_s$ are, respectively, the parts of the elastic energy density which depend on the bulk modulus $K$ and the shear modulus $\mu$. Explicitly

$$W_d = \tfrac{1}{2}K(e_1 + e_2 + e_3)^2 = \frac{1}{18K}(p_1 + p_2 + p_3)^2$$
$$W_s = \mu[e_1^2 + e_2^2 + e_3^2 - \tfrac{1}{3}(e_1 + e_2 + e_3)^2]$$
$$= \frac{1}{4\mu}[p_1^2 + p_2^2 + p_3^2 - \tfrac{1}{3}(p_1 + p_2 + p_3)^2]$$

where $e_1$, $e_2$, $e_3$ are the principal strains and $p_1$, $p_2$, $p_3$ are the principal stresses.

Now force $n$ defects into the strained cube. The work necessary is $n\epsilon$, where $\epsilon$ is the heat of solution per defect, plus $nE_{int}$ where $E_{int}$ is the interaction energy we are looking for. (We suppose for the moment that the defects produce no change of lattice parameter, and that their concentration is small enough for us to neglect any interaction between them.) The total energy of the cube is now

$$E = W_d + W_s + n\epsilon + nE_{int}. \tag{1.61}$$

Start again with the undeformed, defect-free cube. Insert the defects. The energy is $n\epsilon$. Force the cube into the container. Since for fixed strain $W_d$ is proportional to $K$, and $W_s$ to $\mu$, the total energy of the cube is now

$$E = \frac{K + \Delta K}{K} W_d + \frac{\mu + \Delta\mu}{\mu} W_s + n\epsilon \tag{1.62}$$

where $\Delta K$, $\Delta\mu$ are the changes in elastic constants produced by the presence of the defects. If $\Omega$ is the atomic volume and $c$ is the atomic fraction of defects we may write

$$\Delta K = \frac{dK}{dc}n\Omega, \quad \Delta\mu = \frac{d\mu}{dc}n\Omega.$$

The expressions (1.61) and (1.62) refer to the same physical state arrived at in two different ways. Consequently they must be equal, and comparison gives

$$E_{int} = \frac{1}{K}\frac{dK}{dc}W_d\Omega + \frac{1}{\mu}\frac{d\mu}{dc}W_s\Omega. \tag{1.63}$$

The elastic sphere model gives

$$\frac{1}{K}\frac{dK}{dc} = \left[\frac{1+\sigma}{3(1-\sigma)} + \frac{K}{K'-K}\right]^{-1}, \quad \frac{1}{\mu}\frac{d\mu}{dc} = \left[\frac{2}{15}\frac{4-5\sigma}{1-\sigma} + \frac{\mu}{\mu'-\mu}\right]^{-1}$$

where $K$, $\mu$ refer to the matrix and $K'$, $\mu'$ to the inhomogeneous sphere, and $\sigma$ is the Poisson's ratio of the matrix. These expressions suggest that 1 per cent of very hard or very soft atoms will respectively increase or decrease either $K$ or $\mu$ by about 2 per cent.

It is not hard to justify the following recipe for calculating the analogue of (1.63) for an anisotropic material: write out the energy density in terms of the strain components and the elastic stiffness $c_{ij}$ and replaced each $c_{ij}$ by $\Omega dc_{ij}/dc$. Alternatively, write out the energy density in terms of the stresses and the elastic compliances $s_{ij}$ and replace each $s_{ij}$ by $-\Omega ds_{ij}/dc$.

For the model of a vacancy as a spherical hole pulled in by surface tension the interaction energy is the sum of (1.56), (1.63), plus a further term proportional to $p^2$, arising from direct interaction between applied stress and surface energy, which merely adds to the first term of (1.63).

The interaction between point defects and dislocations can be studied by annealing experiments in which the rate at which defects migrate to dislocations is measured. If the interaction law is (1.56) (simple misfit) the number of defects which have reached the dislocation at time $t$ after the beginning of the anneal is proportional to $t^{2/3}$. This is the Cottrell–Bilby (1949) law which has been well verified for, for example, interstitial carbon in iron. If the interaction is given by (1.63) (inhomogeneity but no misfit) the interaction energy is proportional to $r^{-2}$ since the dislocation stresses are proportional to $r^{-1}$. This leads to a $t^{1/2}$ law in place of the $t^{2/3}$ law. Wintenberger's (1957) measurements on the annealing of vacancies in aluminium follow a $t^{1/2}$ law, which suggests that the interaction (1.63) is more important than the interaction (1.56).

According to the isotropic misfitting sphere model the stresses round a point defect are proportional to $r^{-3}$, so that there is an inhomogeneity interaction energy (1.63) proportional to $r^{-6}$ between two point defects. Because the dilatation is zero there is no misfit interaction (1.56). (We ignore the unimportant image correction.) If the material is not isotropic the dilatation is no longer zero and there is an $r^{-3}$ misfit interaction between two defects. If the crystal is

cubic and the defects have cubic symmetry the interaction can still be
calculated from (1.56). When the departure from isotropy is small
the dilatation due to a defect is (Eshelby, 1956, with an error of sign)

$$e = - \frac{\Delta V K}{c_{11}^2} d \frac{x^4 + y^4 + z^4 - \frac{3}{5} r^4}{r^7}$$

where $\Delta V$ is the total volume change it produces, $K$ is the bulk
modulus, and $d = c_{11} - c_{12} - 2c_{44}$ is a combination of the stiffness
constants $c_{ij}$ which vanishes for isotropy. To find the elastic field of a
defect in a crystal which is not nearly isotropic one must know the
elastic Green's function, that is, the displacement due to a concentrated force, and this cannot be written in finite form. Mann et al. (1961)
have calculated it numerically for copper.

## CHAPTER 2

# CRYSTAL DISLOCATIONS

### by F. C. FRANK AND J. W. STEEDS†

## 2.1 INTRODUCTION

Before direct observations could be made on dislocations there were four main classes of experimental evidence pointing to the existence of a class of crystal imperfections which we now analyse in terms of the dislocation concept. The earliest body of evidence was that from X-ray diffraction: within a year or two of the discovery of this phenomenon by Friedrich, Knipping, and von Laue in 1912, it became evident that the diffraction intensities from most real crystals were very different from those one would calculate on the assumption of ideal crystal structures. A rather ill-defined concept of mosaic structure was employed for a generation to account in general terms for these discrepancies. We have ultimately learnt that their cause is sometimes well described by the term 'mosaic structure', sometimes not, but in either case we are able to describe the inferred departure from the ideal structure in some detail in terms of dislocations. The next body of evidence was concerned with the mechanical properties of solids, which, to put it briefly, are a thousand times weaker than ideal crystals. The fact became increasingly evident as people acquired confidence in the Born type of model for the ideal crystal, which accounts very well for the elastic or thermal properties, but indicates plastic strengths vastly in excess of those of technical materials. It was to explain these discrepancies that Taylor and Orowan independently put forward the concept of crystal dislocations in 1934. Related ideas had been employed by Prandtl (1928), Dehlinger (1929), Polanyi (1934), and Smekal (1929) at earlier dates, but the modern development starts very clearly from these two papers in 1934. It was followed by fifteen years or so of theoretical development, with no direct experimental observation of dislocations, leading to a cleavage between metal physicists who employed the dislocation

† Dr Frank is Professor of Physics and Dr Steeds is Lecturer in Electron Microscopy at the H. H. Wills Physics Laboratory, University of Bristol.

concept and many metallurgists, who regarded it as a complex fiction introduced to rescue an unsuccessful theory. A third body of evidence was latent in the literature – a very large discrepancy between observed rates of crystal growth and rates calculated on the assumption that crystals were ideal. The very large quantitative discrepancy was obscured by a qualitative similarity between observed and calculated laws of dependence of growth rate on super-saturation. Once recognised, it became another experimental fact attributable to the presence of dislocations in real crystals and very soon led to the clearest visual evidence that they were present. By this time, the study of point defects had also given compelling evidence that the crystal surface was not the only place where the number of crystal lattice sites could alter – for example, the rate of formation of $F$-centres in alkali halides under irradiation was too great to suppose that vacancies could only form at the surface.

During the past twenty years several methods of direct observation of dislocations have been explored. Of these, transmission electron microscopy is undoubtedly the most generally useful technique, combining high resolution with great powers of analysis through diffraction contrast. Lang topography, the equivalent technique using X-rays, has been widely applied to the study of materials of relatively low atomic number containing not too many dislocations. The formation of etch pits at dislocations is a simpler process in favourable cases, working at much the same resolution as X-ray topography but only revealing dislocations intersecting one of a few low index crystal planes. Techniques requiring the decoration of dislocations, which have been developed for example by Amelinckx (1957) (alkali halides) and Mitchell (1957) (photosensitive materials), are of more limited application. As a result of the widespread applica-tion of these various observational techniques we now have precise evidence about the actual arrangements of dislocations in many real crystals.

Dislocation theory falls into parts of various status. There are some conclusions which are exact, because they are purely geometrical or topological. There are some parts which though approximate are nevertheless very reliable (for example, the conclusion that disloca-tions should be absent from a crystal in thermodynamic equilibrium – thermodynamic properties of dislocations have only been roughly estimated but nevertheless the error could not be large enough to reverse the conclusion). And then, naturally, many of the questions of current interest are those for which one cannot be sure of the reliability of approximations made and must draw on all the resources of theory, observation, and guesswork in the attempt to make further progress. The theory of work hardening of metals is the classical example of this process.

The most familiar picture of a dislocation is Taylor's picture of the edge dislocation (fig. 2.1). This picture has to be completed three-dimensionally (the dislocation line being perpendicular to the plane of the drawing) at the edge of the incomplete lattice plane. Though

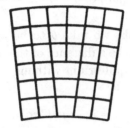

Fig. 2.1. Edge dislocation

easiest to draw, the edge dislocation is not the simplest one, which is rather the screw dislocation, the significance of which was first pointed out by Burgers in 1939. To describe the state of strain at a screw dislocation (fig. 2.2) we picture a hollow cylinder, cut along a radial surface at which the two cut faces are given an axial relative displacement of amount $b$. Then every element of a cylindrical shell of thickness $dr$ at radius $r$ in this cylinder has a shear strain

$$\gamma = b/2\pi r.$$

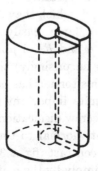

Fig. 2.2. Screw dislocation. **b** is an upward vector parallel to the axis of the cylinder, of magnitude equal to the step height

The cylindrical surfaces of such a shell are free from stress, so that each shell is in equilibrium within itself provided the stresses are matched by suitable stresses at the end surfaces: we assume for the moment that these are present. The shear involves an energy density

$$\tfrac{1}{2}\mu\gamma^2 = \tfrac{1}{2}\mu b^2/4\pi^2 r^2$$

where $\mu$ is the shear modulus. Hence the elastic energy per unit length in a cylinder of internal radius $r_0$ and external radius $R$ is

$$E_0 = \int_{r_0}^{R} (\tfrac{1}{2}\mu b^2/4\pi^2 r^2)\, 2\pi r\, \mathrm{d}r = \frac{\mu b^2}{4\pi}\ln\frac{R}{r_0}.$$

In application to crystals, we take $b$ to be an interatomic spacing and the inner radius $r_0$ to be a distance of the same order. This condition is not perfectly in equilibrium without applied tractions at the ends, having the magnitude

$$\tau = \mu\gamma = \mu b/2\pi r$$

producing a torque which is found by integration to be

$$\tfrac{1}{2}\mu b(R^2 - r_0^2) \sim \tfrac{1}{2}\mu b R^2.$$

If these tractions are absent there will be additional displacements produced by negative tractions of the same magnitude. Except within a few diameters of the ends these displacements are to be described as a simple torsion produced by the given torque. This amounts to

$$\mathrm{d}\theta/\mathrm{d}z = b/\pi R^2.$$

This torsion is not of much importance when $R$ is large, but in the case of a thin metal 'whisker', which may have a screw dislocation along its axis, it becomes a large, easily observed quantity (e.g. about 45° per cm if $b = 2.5$ Å and $R = 1$ $\mu$m). It has been shown by Eshelby (1953) that (rather surprisingly) the screw dislocation will be stable in this position. This torsion has been observed in whiskers by several workers. It is quite important in transmission electron microscopy since it reveals screw dislocations (or dislocations with a significant screw component) normal to the foil under observation by a characteristic form of contrast (fig. 2.3). The strain field in this case can be obtained quite simply by the method of images employed in electrostatic theory.

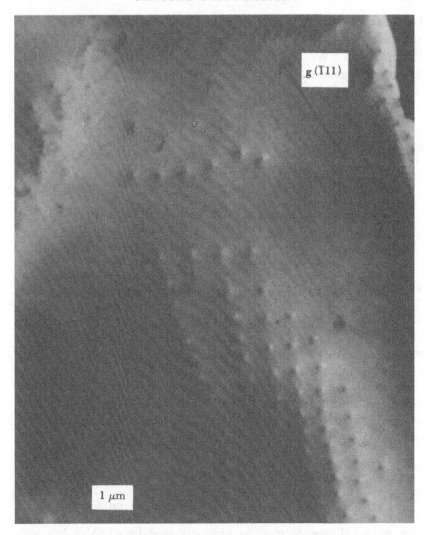

Fig. 2.3. Approximately square array of 60° dislocations viewed end-on in a thin platinum foil; the contrast arises from surface relaxation effects associated with the screw component. $\mathbf{g}$ corresponding to reflection from the plane ($\bar{1}11$) is the diffraction vector used in forming this bright-field electron micrograph. (Tunstall *et al.*, 1964)

## 2.2  ONE-DIMENSIONAL MODELS

The first purpose of introducing the idea of crystal dislocations was to explain plastic deformation. Simple consideration of fig. 2.1 indicates that the dislocation ought to be relatively easily displaced through the crystal, producing a plastic deformation. However the model of continuum elasticity which we have employed in the discussion of the screw dislocation is of little value, directly, for discussing dislocation mobility. We therefore study some one-dimensional models having properties related to those of dislocations:

1. A heavy rope lying on corrugated iron (fig. 2.4);

2. A sinusoidal surface on which a series of weights is put, the weights being connected by springs (a model considered by Frenkel and Kontorova (1938, 1939)); see fig. 2.5.

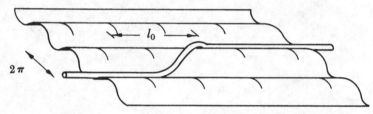

Fig. 2.4. Mechanical dislocation model corresponding to equation (i) in section 2.2

Fig. 2.5. Mechanical dislocation model corresponding to equation (ii) in section 2.2

If $\alpha$ is either the tension in the rope or the modulus of the springs, and $W$ is a measure of the amplitude of the sinusoidal potential, then we have for the displacement $u$ from an equilibrium position (the substrate period in $u$ being $2\pi$) the following equations:

(i) $\alpha \dfrac{\mathrm{d}^2 u}{\mathrm{d}x^2} = W \sin u$     ($x$, co-ordinate along the rope)

(ii) $\alpha(u_{m+1} - u_m) - \alpha(u_m - u_{m-1}) = W \sin u_m$;     ($m = 1, 2, 3 \ldots$).

The second difference may be approximated by $\mathrm{d}^2 u/\mathrm{d}x^2$ ($x = md$, where $d$ is the equilibrium spacing of the weights), in which case the two equations are identical. These equations are of the form

$$\frac{\mathrm{d}^2 u}{\mathrm{d}x^2} = y \sin u \qquad y = W/\alpha.$$

The general solution is expressed by elliptic integrals (representing sequences of 'dislocations') but we may be content with the solution representing a single dislocation:

$$\tan (u/4) = \exp (-y^{1/2}x).$$

The effective length of the dislocation may be defined as

$$l_0 = \frac{2\pi}{(\mathrm{d}u/\mathrm{d}x)_{\max}} = \pi \sqrt{\frac{\alpha}{W}}.$$

If we use this model to represent a monolayer of atoms in a crystal surface (fig. 2.6), and choose the constants of the equations to agree with interatomic forces of the form $-ar^{-7}+br^{-13}$, we find that $l_0$ is about 7 interatomic spacings.

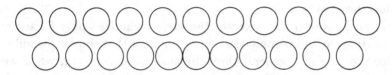

Fig. 2.6. Monolayer (upper) of atoms on a surface (lower)

A discussion of the motion of dislocations according to this model has recently been presented by Weiner (1970). Either version of our model, corresponding to fig. 2.4 or 2.5, (or alternative equivalent models) readily exhibits the effect of mobility of the dislocation on its energy and strain field. To include the possibility of moving, the foregoing equation is to be replaced by

$$\alpha \frac{\partial^2 u}{\partial x^2} - \rho \frac{\partial^2 u}{\partial t^2} = W \sin u$$

which becomes

$$\frac{\partial^2 u}{\partial x^2} - \frac{1}{c^2}\frac{\partial^2 u}{\partial t^2} = y \sin u$$

with $c = (\alpha/\rho)^{1/2}$ the velocity of sound in the rope, or chain of springs and masses, in the absence of the periodic substrate; $\rho = M/d$, where $M$ is the particle mass. For the case of steady motion of a dislocation with velocity $v$ we may solve this equation by introducing a new variable

$$X = \frac{x - vt}{\sqrt{(1 - v^2/c^2)}}.$$

This reduces the equation to the same form as we had for the static case, the effective length being reduced by a 'Lorentz contraction' factor $\sqrt{(1-\beta^2)}$ with $\beta = v/c$. One can also calculate the energy

$$E = \tfrac{1}{2}E_0\left[\frac{1-\beta^2}{\sqrt{(1-\beta^2)}} + \frac{1}{\sqrt{(1-\beta^2)}} + \frac{\beta^2}{\sqrt{(1-\beta^2)}}\right] = \frac{E_0}{\sqrt{(1-\beta^2)}}$$

consisting of three parts, the first two being the potential energy of the weights or the rope brought out of their equilibrium position at the bottom of groove or well and the work done against tension in rope or springs, and the third part the proper kinetic energy. We have a contracted strain field and an increased energy when the dislocation moves.

It can be shown that exactly the same formulae apply for screw dislocations in an isotropic elastic medium and that similar formulae cover the case of a general three-dimensional dislocation, when we have the complication that more than one speed of sound is involved. As the dislocation approaches the speed of sound its energy approaches infinity. It is believed that in ordinary slip processes dislocations do not move faster than $c/10$, but they may move considerably faster in some processes of twinning or martensitic transformation and a fully developed theory of these processes may need to take the relativistic increase of energy into account. For low velocity processes we may usefully derive the result that the effective mass of a dislocation line is equal to its rest energy divided by $c^2$.

## 2.3  PEIERLS–NABARRO MODEL

So far, we have used the linear theory of isotropic elasticity to describe the state of strain of a hollow cylinder. The shear strain $\gamma = b/2\pi r$ will amount to a few per cent if the radial distance $r$ is about two or three times $b$. Crystals will take linearly strains of the order of 2 or 3 per cent. To take non-linearity into account one can adopt the model proposed by Peierls (1940) and further developed by Nabarro (1947), Leibfried and Dietze (1949), Eshelby (1949), and Van der Merwe (1950). We cut the crystal into two pieces, in which we assume linear elasticity theory to be valid, and assume a special type of connection between the two surfaces. If $u_1$ and $u_2$ are the displacements of corresponding points on the two surfaces, the tangential forces are of the form $A\sin(u_2 - u_1)$, the constant $A$ being chosen so that the maximum slope of the graph of the restoring force versus relative displacement corresponds to the rigidity modulus of the material acting in the gap of width $a$.

Leibfried and Dietze (1949) showed that if the gap $a$ between the two surfaces is equal to the distance $b$ between atoms along the slip direction, the strain–stress systems are identical with the strain–stress systems given by the continuous model. The same type of model may be used both for edge dislocations and for screw dislocations, differing only in the direction of the displacements relative to the dislocation line. In general the energy per unit length is

$$E_0 = \mu \frac{b^2}{4\pi K} \ln \frac{KR}{a}$$

where $\mu$ is the rigidity modulus, $K = 1$ for screw dislocations and $K = 1 - \nu$ for edge dislocations ($\nu$, Poisson's ratio). Thus X is between 1 and 0.7 for most materials.

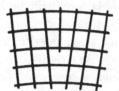 

Fig. 2.7. Two symmetrical positions of an edge dislocation

An important question for the mobility of dislocations is how much the energy fluctuates as the dislocation moves. During its motion, an edge dislocation passes from the symmetrical configuration 1 to the different symmetrical configuration 2 (see fig. 2.7). The Peierls model gives the result that the energy in the first and second configurations are identical, and that as the dislocation moves the energy fluctuation has a maximum between the two symmetrical configurations. This is a consequence of the simple sinusoidal form assumed for the non-linear interaction. Huntington (1955), has discussed this point, and proposed some refinements. Foreman, Jaswon and Wood (1951) considered the case of a more general form of tangential force. According to the Peierls–Nabarro theory the amplitude of the energy fluctuation is

$$\Delta W = \mu \frac{b^2}{2\pi K} \exp\left(-\frac{2\pi a}{Kb}\right)$$

and the corresponding critical shear stress is

$$\sigma = \frac{2\mu}{K} \exp\left(-\frac{2\pi a}{Kb}\right).$$

From this result it is clear that the ratio $a/b$ is a crucial factor in determining the operative slip system(s), and since in the equation above $a$ is the inter-planar spacing we find qualitative agreement with the observed slip systems of f.c.c. metals. However, no form of Peierls–Nabarro model is sufficiently realistic to enable a practical calculation to be made of the force required to lift a dislocation out of its potential valley (the so-called Peierls force) although there have been many attempts to adapt it for this purpose. One clear example of its inadequacy is the $\langle 111 \rangle$ screw dislocation in b.c.c. metals which lies along a three-fold, not two-fold, axis. The consequent atomic adjustment at the core leads to a Peierls force of essentially different character. Another example is the case of dislocations in diamond or sphaleritic structures where the strongly directed bonds invalidate the model. What is clear is that the Peierls force will be high for dislocations lying along closely packed directions, for example $\langle 111 \rangle$ in b.c.c. structure and $\langle 110 \rangle$ in diamond or sphalerite, and will vanish along non-crystallographic directions. As a result, dislocations will tend to align along these hard directions, a conclusion for which experimental evidence can be gathered (fig. 2.8).

In crystals where the Peierls force dominates the mobility of dislocations, Mott and Nabarro (1948), Read (1953), and Seeger (1956) have proposed and developed the idea that motion will occur by the nucleation and motion of kinks along dislocations lying in Peierls troughs (fig. 2.4). This idea, though most plausible, has proved almost impossible to verify owing to the difficulty of either direct observation of such small defects, or of reasonable calculations of the kink energies.

In the absence of more sophisticated methods, relaxation calculations of atomic arrangements at dislocation cores have recently been performed for b.c.c. metals using high speed computers and carefully chosen interatomic potentials. These calculations have produced qualitatively interesting results although they are at present subject to severe uncertainties concerning the number of interacting neighbouring atoms assumed, the total number of atoms considered, the boundary conditions to the crystallite, and the nature of the interatomic potential. Some of the calculations do not even give stable b.c.c. lattices. Atomistic calculations of this form were performed some years ago by Huntington (1955) for dislocations in NaCl using a Born–Mayer interaction potential. Recently these calculations have been extended to a number of different interaction potentials and the Peierls force has been determined.

200 $\mu$m

Fig. 2.8. X-ray topograph of dislocations from a spiral source in silicon. (Authier and Lang, 1964)

## 2.4 EDGE DISLOCATION IN A CYLINDER

One way of creating an edge dislocation is to displace radially a wedge of material generated by two radial planes in a circular cylinder of elastic material (fig. 2.9). On closing and cementing the neighbouring planar surfaces an edge dislocation is formed. The procedure, proposed by Eshelby (1966), allows a particularly clear derivation of the

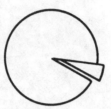

Fig. 2.9. Eshelby's representation of an edge dislocation by displacement of a wedge disclination

displacement and stress fields. However, for the sake of development to the more general case in section 2.6, we prefer to use a more formal approach based on the hollow cylinder of fig. 2.2 deformed by a displacement $b$ perpendicular to its axis. We require the appropriate solution of the equilibrium and compatibility equations of elasticity theory. For an infinite straight edge dislocation in the isotropic approximation we may assume that all displacements lie in a plane perpendicular to the dislocation (plane strain). Under these circumstances, for a dislocation along the $z$-axis of a right-handed Cartesian co-ordinate system, the equations become

*Equilibrium*

$$\frac{\partial \sigma_{xx}}{\partial x} + \frac{\partial \sigma_{xy}}{\partial y} = 0 \quad \text{and} \quad \frac{\partial \sigma_{yx}}{\partial x} + \frac{\partial \sigma_{yy}}{\partial y} = 0$$

*Compatibility*

$$\frac{\partial^2 e_{xx}}{\partial y^2} + \frac{\partial^2 e_{yy}}{\partial x^2} = 2 \frac{\partial^2 e_{xy}}{\partial x \, \partial y} \tag{2.1}$$

where the strain tensor $e_{ij}$ is derived from the displacements $u$ according to $e_{ij} = \frac{1}{2}(\partial u_i/\partial x_j + \partial u_j/\partial x_i)$. Note that, owing to the plane strain condition, $z$ components and differentials with respect to $z$ vanish.

By making the substitution

$$\sigma_{xx} = \frac{\partial^2 \chi}{\partial y^2} \quad \sigma_{yy} = \frac{\partial^2 \chi}{\partial x^2} \quad \sigma_{xy} = -\frac{\partial^2 \chi}{\partial x \, \partial y} \tag{2.2}$$

the equilibrium equations are automatically satisfied. $\chi$ is an Airy function. Substitution in the compatibility equation yields the prescription for $\chi$

$$\nabla^4 \chi = 0.$$

With the assumption $\chi = R(r) . \theta(\theta)$, and adopting as a trial solution

$\theta(\theta) = \sin\theta$, one has

$$\chi = A\left(r\ln r + \frac{r_0^2}{2r} - \frac{r^3}{2R^2}\right)\sin\theta$$

where $A$ is a constant determined by the relative displacement, $b$, of the cut surfaces and the shear modulus and is equal to $-\mu b/2\pi(1-\nu)$. $r_0$ and $R$ are constants chosen to give zero stresses on the inner and outer cylindrical surfaces and $\theta$ is measured from the slip plane. The essential term describing the dislocation is the logarithmic term. As a result, one has a compression in the half-crystal which contains the extra half-plane of the dislocation and a dilatation in the other half-crystal, the two half-crystals being separated by the slip plane of the dislocation, though the situation is not always as simple as this when elastic anisotropy is taken into account (see section 2.6).

Consideration of the stresses at the ends shows that an edge dislocation along the axis should not produce any bending of the cylinder analogous to the torsion which occurs with screw dislocations.

The resulting stress field (omitting the surface terms) is most simply described in polar co-ordinates:

$$\sigma_{rr} = \sigma_{\theta\theta} = \frac{-\mu b}{2\pi(1-\nu)} \cdot \frac{\sin\theta}{r}$$

$$\sigma_{\theta r} = \frac{\mu b}{2\pi(1-\nu)} \cdot \frac{\cos\theta}{r}$$

$$\sigma_{zz} = \frac{-2\nu\mu b}{2\pi(1-\nu)} \cdot \frac{\sin\theta}{r}.$$

However, the most important stress is the shear stress, $\sigma_{xy}$, on planes parallel to the glide plane, $\theta = 0$. This is expressed by

$$\sigma_{xy} = \frac{\mu b}{2\pi(1-\nu)} \cdot \frac{\cos\theta\cos 2\theta}{r}$$

illustrated in fig. 2.10 (the $\sigma_{xx}$ and $\sigma_{yy}$ may be derived directly from this – see section 2.6). Of especial importance is the existence of planes of zero stress at 45°, 90°, etc., with sectors of reversed stress, leading to a position of stable equilibrium for a second similar dislocation in the direction $\theta = \pi/2$.

The energy per unit length of the edge dislocation (apart from

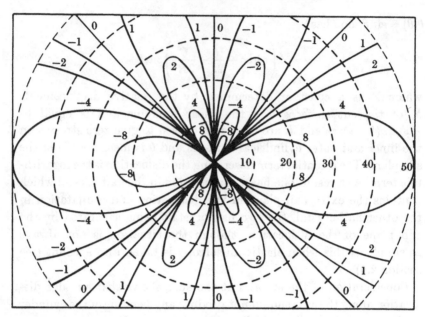

Fig. 2.10. The $\sigma_{xy}$ field of an edge dislocation; unit of stress $= \mu/400\,(1-\nu)$, unit of distance $b$

surface terms) is

$$E_0 = \frac{\mu b^2}{4\pi(1-\nu)} \ln\left(\frac{R}{r_0}\right).$$

The inner hole of radius $r_0$ is, thus far, only a device to eliminate the singularity. The appropriate value of $r_0$ for the energy formula can be estimated in various ways. According to the elastic theory, the energy density varies as $r^{-2}$; but Hooke's law fails when $r < {\sim}5b$ where the strain is about 3 per cent. It fails in such a manner that the energy density increases less rapidly. We may call the energy density within this radius the core energy. The Peierls–Nabarro model gives us one value for it. Bragg (1948) estimated what should be an upper limit by assuming obedience to Hooke's law till the energy density reaches the latent heat of melting, and a constant energy density in the region closer to the centre. In the case of atomistic calculations the core energy has generally been evaluated directly but there is reason to believe that these values are considerably less reliable than the atomic positions deduced. Some experimental estimates can be obtained from 'observed' values of the surface tension of small-angle intercrystalline boundaries, which, as we shall see, are really systems of dislocations. By these various methods we find that the core energy can be absorbed into the elastic expression by putting $r_0$ equal to about 1 or 2 times $b$.

Let us now consider the magnitude of $R$. Our cylinder is a rather artificial example, but we may solve the elastic problem for a variety of other external geometries, always obtaining a dominant energy term of the same form, with $R$ of the order of magnitude of a macroscopic dimension of the situation – for example, the nearest distance between the dislocation and a free surface, the half-distance between a parallel pair of dislocations of opposite sign, or the radius of a dislocation loop. In these cases the stress field nearer to the dislocation than $R$ is closely similar to that of an isolated dislocation in an infinite body, falling to zero at distances large compared with $R$. As a rule, no great precision is required in estimating $R$. If we say $R$ lies between 1 and 1000 $\mu$m, and $r_0$ between 0.5 and 5 Å, the range of variation of $\ln(R/r_0)$ is about a factor 2. It can often be treated as a constant without serious error and its magnitude approximately cancels the divisor $4\pi K$ ($K$ lying between about 0.7 and 1 according as the dislocation is screw or edge). Then we have $E_0 \sim \mu b^2$: or the energy per atomic length along the dislocation line is $\sim \mu b^3$. For elementary dislocations in metals this is of the order of a few eV (see also Hirth and Lothe, 1968).

## 2.5 MORE GENERAL DISLOCATION LINES

Figure 2.11 is a picture, deriving from J. M. Burgers (1939), which introduced much more generality and flexibility into our thinking about dislocations. One should visualise the black dots as lying in front of the plane of the picture, with essentially similar patterns repeated in further planes in front of this, and the open circles as lying behind the plane of the picture, with essentially similar patterns repeated in further planes behind this, Then we have an edge dislocation line leading forwards, out of the plane of the picture, connecting with a screw dislocation line leading to the right, in the plane of the picture. This screw dislocation lies in the slip plane of the edge dislocation, and the same plane can serve as a slip plane for the screw. From this we can proceed to the representation of a dislocation line which is of mixed edge and screw character and lies along an arbitrary path in the glide plane – first, as a stepped line of edge and screw portions, then, in the limit, as a line of arbitrary orientation or

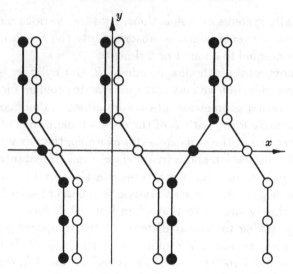

Fig. 2.11. Burgers' representation of the junction of an edge and a screw dislocation

continuous curvature. The centre of the dislocation line can be of more continuous curvature than the discrete lattice structure of the crystal because the dislocation is several atoms wide at the core.

To the approximation of isotropic elastic theory, we obtain the stress and strain fields of a dislocation in arbitrary orientation by superposing those of edge and screw dislocation in appropriate proportions (given by $\sin\psi$ and $\cos\psi$, where $\psi$ is the angle between slip direction and dislocation line). Such superposition is always permissible when Hooke's law is obeyed. In the present case, since the stresses of edge and screw dislocations are orthogonal, we can also add the energies (in proportions $\sin^2\psi$ and $\cos^2\psi$). Since, furthermore, the energies of edge and screw dislocations are not very different (roughly in the ratio 4 to 3) we can often disregard the variation of this orientation-dependent factor as well as the variation of the factor $\ln(R/r_0)$ and proceed as though dislocation lines had a constant energy per unit length. This is the basis of the method introduced by Mott and Nabarro (1948) for the approximate calculation of the elastic equilibrium of dislocation lines in complex configurations: in this method we regard them as dislocation lines having a constant line tension equal to their energy per unit length.

An example of a general dislocation line is illustrated in fig. 2.12. The dislocation line, which is edge on the surface plane $ABCD$ and

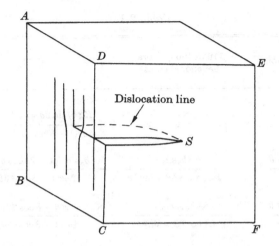

Fig. 2.12. A curved dislocation line

screw on the surface plane $CDEF$ transformed into a helicoid, is continuously changing in character in the intermediate region.

It is intuitively obvious that an applied shear stress will tend to expand the slipped area in all directions. We show later that the effective force per unit length on the dislocation in its glide plane is proportional to the magnitude of $b$ and to the shear stress in the glide plane resolved in the slip direction. That this is true on the average is very simply shown by considering the work done when a dislocation moves right across a crystal.

## 2.6 STRAIGHT DISLOCATIONS IN ANISOTROPIC ELASTICITY

To obtain results equivalent to those quoted in sections 2.1 and 2.4 in the case of anisotropic elasticity we follow the approach of section 2.4 and pay no attention to the boundary conditions acting on the anisotropic space. Although the edge and screw components of a dislocation can still be treated separately on account of linearity, and displacements or stresses can be added together, the separation of terms which one has in the isotropic case is lost. Owing to this departure from plane strain conditions for an edge dislocation, and the variation of strain in the cylindrical shell around a screw dislocation, it is necessary to supplement both the equilibrium and compatibility conditions by further equations and then to look for appropriate

## *Table* 2.1

| Stress components | Isotropic | $\langle 110\rangle/\langle 100\rangle$/several hexagonal cases | $\langle 111\rangle$ |
|---|---|---|---|
| *Screw* | | | |
| $\sigma_{xy}$ | $0$ | $0$ | $-\dfrac{K_3^* b_{\mathrm{s}}\sin 2\theta\sin 3\theta}{4\pi r(1-\delta\cos^2 3\theta)}$ |
| $\sigma_{xz}$ | $-\dfrac{\mu b_{\mathrm{s}}\sin\theta}{2\pi r}$ | $-\dfrac{K_3 b_{\mathrm{s}}\sin\theta}{2\pi r(A\cos^2\theta+\sin^2\theta)}$ | $\dfrac{K_3' b_{\mathrm{s}}\sin\theta(1+2\delta\cos\theta\cos 3\theta)}{2\pi r(1-\delta\cos^2 3\theta)}$ |
| $\sigma_{yz}$ | $\dfrac{\mu b_{\mathrm{s}}\cos\theta}{2\pi r}$ | $\dfrac{K_3 b_{\mathrm{s}} A\cos\theta}{2\pi r(A\cos^2\theta+\sin^2\theta)}$ | $\dfrac{K_3' b_{\mathrm{s}}(\cos\theta-\delta\cos 2\theta\cos 3\theta)}{2\pi r(1-\delta\cos^2 3\theta)}$ |
| $u_z$ | $\dfrac{b_{\mathrm{s}}\theta}{2\pi}$ | $\dfrac{b_{\mathrm{s}}}{2\pi}\tan^{-1}\dfrac{\tan\theta}{\sqrt{A}}$ | $\dfrac{b_{\mathrm{s}}}{6\pi}\tan^{-1}\left(\dfrac{\tan 3\theta}{\sqrt{(1-\delta)}}\right)$ |
| *Edge* | | | |
| $\sigma_{xy}$ | $\dfrac{\mu b_{\mathrm{e}}\cos\theta\cos 2\theta}{2\pi(1-\nu)\,r}$ | $\dfrac{K_1\,\mathbf{b}_{\mathrm{e}}\cdot\mathbf{r}(\cos^2\theta-\lambda^2\sin^2\theta)}{2\pi r^2(\cos^4\theta+\lambda^4\sin^4\theta}$ $-\dfrac{\Lambda}{2}\sin^2 2\theta)$ | $\dfrac{K_1' b_{\mathrm{e}}\cos\theta(\cos 2\theta-\delta\cos\theta\cos 3\theta)}{2\pi r(1-\delta\cos^2 3\theta)}$ |

$$A=\frac{2c_{44}}{c_{11}-c_{12}},\quad K_3=\frac{c_{44}}{\sqrt{A}},\quad K_3'=\frac{c_{44}}{\sqrt{(1-\delta)}},\quad K_3^*=\frac{(3\sqrt 2)c_{44}\delta}{(A-1)\sqrt{(1-\delta)}}$$

$\lambda,\Lambda,\delta,K_1,K_1'$ are more complicated expressions involving the elastic constants. Values of $\lambda$ (for {111} slip), $\Lambda$ (for {100} slip), and $\delta$ are given in table 2.2 for some typical materials. In the isotropic limit, both $K_1'$ and $K_1$ tend to $\mu/(1-\nu)$: $\lambda,A\to 1$: $\delta\to 0$: $\Lambda\to -1$.

solutions. The additional equations are

*Equilibrium*

$$\frac{\partial\sigma_{zx}}{\partial x}+\frac{\partial\sigma_{zy}}{\partial y}=0$$

*Compatibility*

$$\frac{\partial}{\partial x}\left(-\frac{\partial e_{yz}}{\partial x}+\frac{\partial e_{xz}}{\partial y}\right)=0. \tag{2.3}$$

Steeds (1973) has shown that solutions may be most easily found by introducing a second stress function $\Phi$, in addition to $\chi$, such that

$$\sigma_{xz} = -\frac{\partial \Phi}{\partial y} \quad \sigma_{yz} = \frac{\partial \Phi}{\partial x} \tag{2.4}$$

thereby satisfying the additional equilibrium condition. The generalised form of Hooke's law may be expressed

$$e_{ij} = s_{ijkl}\,\sigma_{kl} \quad \text{(summation convention, } ijkl = 1, 2, 3) \tag{2.5}$$

where $ij/kl$ may be coupled together in $s_{ijkl}$ and reduced as follows: $ii \to i$, $12 \to 6$, $13 \to 5$, $23 \to 4$; $s_{ij}$ are the elastic compliances. Substituting first (2.2) and (2.4) into (2.5) and then the resulting equations into the compatibility equations, we obtain (complicated) coupled fourth order differential equations of first degree for the stress functions $\chi$ and $\Phi$. Their solution is the central problem of anisotropic elasticity theory. In general the solution leads to a sextic equation. However, owing to the symmetry properties of the dislocation axes, there are several cases where this sextic simplifies to a bicubic or quadratic equation. In these cases analytic solutions can be found for the stress fields, displacements and energies: some examples are shown in table 2.1. The headings of the third and fourth columns in this table indicate the direction of the dislocation axis, and it is also important to know the orientation of the other axes. For the $\langle 110 \rangle$ dislocation the $x$-axis is $[\bar{1}10]$ and the $y$-axis $[001]$, while for the $\langle 100 \rangle$ dislocation the $x$- and $y$-axes are the other cube axes. For the $\langle 111 \rangle$ dislocation the $x$-axis is $[11\bar{2}]$ and the $y$-axis is $[\bar{1}10]$.

Having obtained one 'edge-like' (i.e. $\sigma_{xx}$, $\sigma_{xy}$ or $\sigma_{yy}$) stress field, the others may be derived from it. This step is the equivalent of the stress function method (equation (2.2)) but it is now realised through the expression for the dislocation energy which Foreman (1955) found to have the general form

$$E_0 = \frac{K_{ij} b_i b_j}{4\pi} \ln\left(\frac{R}{r_0}\right) \quad \text{(summation conversion)}$$

where for the $\langle 110 \rangle / \langle 100 \rangle$ dislocation

$$K_{11} = K_1 \quad K_{22} = K_1/\lambda^2 \quad K_{33} = K_3 \quad K_{ij} = 0 \text{ if } i \neq j.$$

For the $\langle 111 \rangle$ dislocation $K_{11} = K_{22} = K_1'$, $K_{33} = K_3'$, $K_{ij} = 0$ if $i \neq j$. Expressions for $K_{ij}$ and $\lambda$ in terms of the elastic constants (see table 2.1) can be found in Steeds (1973). Taking radial cuts in a cylinder of anisotropic elastic material and displacing the cut surfaces to make

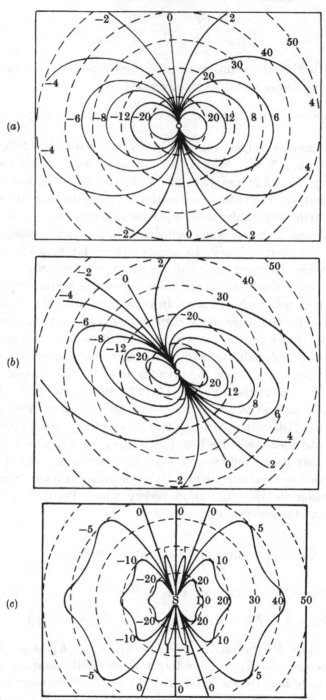

an edge dislocation we find expressions for $\sigma_{rr}$, $\sigma_{r\theta}$ and $\sigma_{\theta\theta}$ in terms of $K_{ij}$. From these we obtain

$$\sigma_{xx} = \sigma_{xy}\cot\theta - K_1 b_1/2\pi r\sin\theta$$

$$\sigma_{yy} = \sigma_{xy}\tan\theta + K_{22} b_2/2\pi r\cos\theta.$$

These equations are valid for all the cases considered here, i.e. $\langle 110\rangle$ $\langle 100\rangle$ $\langle 111\rangle$ edge and screw components. Finally, as will be mentioned in section 2.12, the stress fields $\sigma^e_{13}$ and $\sigma^e_{23}$ can be derived from the table using the relations

$$b_e\,\sigma^s_{12} = b_s\,\sigma^e_{23}, \quad b_e\,\sigma^s_{11} = b_s\,\sigma^e_{13} \qquad (2.6)$$

so that all the stress fields are known for the cases considered.

Some results are illustrated in fig. 2.13 for the $\sigma^s_{yz}$ field. Examination of the appropriate equation in table 2.1 shows that for $0 < \delta < \frac{1}{3}$ there are no additional lines of zero stress in the anisotropic $\sigma^s_{yz}$ field. However, for $\frac{1}{3} < \delta < 1$, a situation not uncommon in practice (table 2.2), two lines of zero stress occur for $\cos^2\theta = [5 - (1 + 8/\delta)^{1/2}]/8$

*Table 2.2*

| Material | $A$ | $\lambda$ | $\Lambda$ | $\delta$ |
|---|---|---|---|---|
| Li (195 °K) | 8.73 | 1.13 | 0.62 | 0.52 |
| Fe | 2.4 | 1.07 | −0.07 | 0.11 |
| W | 1.00 | 1.00 | −1.00 | 0.00 |
| Nb | 0.51 | 0.97 | −2.47 | 0.08 |
| Au–50at.%Cd | 11.7 | 1.09 | 0.74 | 0.62 |
| Cu | 3.2 | 1.07 | 0.18 | 0.20 |
| Al | 1.01 | 1.01 | −0.72 | 0.01 |
| KCl | 0.37 | 0.93 | −2.97 | 0.14 |

(see case of lithium in fig. 2.13). Likewise, the $\sigma^s_{xy}$ field exhibits extra lines of zero stress if $\frac{8}{9} < \delta < 1$ at $\cos^2\theta = [3 \pm (9 - 8/\delta)^{1/2}]/8$, though this implies a very high degree of elastic anisotropy. Another interesting conclusion from the stress fields in table 2.1 is that the sign of the $\sigma^s_{xy}$ field depends on whether $A$ is greater than or less than unity. Finally, the limiting form of these stress fields at the crystal stability limit $s_{11} \to 2s_{12}$ (i.e. $c_{11} \to c_{12}$) can be deduced since $\Lambda \to 1$ and $\delta \to 1$.

Fig. 2.13. The $\sigma_{xy}$ field of a screw dislocation in various materials: (a) aluminium, (b) copper, (c) lithium. The unit of stress is $10^9/\pi$ in (a) and (b), $10^8/2\pi$ in (c); the unit of distance is $b$

It is found that there are important differences between the isotropic and anisotropic dilatational fields. For a screw dislocation isotropic elasticity predicts vanishing dilatation. This result also follows in anisotropic elasticity for the $\langle 110 \rangle / \langle 100 \rangle$ screw dislocation, but the $\langle 111 \rangle$ screw has a finite dilatation

$$\Delta_s \propto \frac{b_s \, \delta \sin 3\theta}{2\pi r \, (1 - \delta)^{1/2} (1 - \delta \cos^2 3\theta)}.$$

Moreover, in sufficiently anisotropic crystals extra sectors occur in the edge dilatational field. Let us take the hypothetical case of a $\langle 100 \rangle \{001\}$ edge dislocation. The dilatation would be

$$\Delta_e = \frac{\sigma_{xx} + \sigma_{yy}}{c_{11} + c_{12}}$$

where $c_{11}$ and $c_{12}$ are the elastic stiffness constants. From table 2.1 we have

$$\Delta_e = - \frac{K_1 \, b_e \sin \theta \, [(\Lambda + 1) \sin^2 \theta - \Lambda]}{2\pi r (c_{11} + c_{12})}$$

i.e. extra zeros occur for the interval $1 > \Lambda > 0$. At the crystal stability limit ($\Lambda = 1$) $\theta = \pm \pi/4$, $\pm 3\pi/4$. The form of dilatational field for an example in this interval ($\Lambda = \frac{1}{2}$) is shown in fig. 2.14.

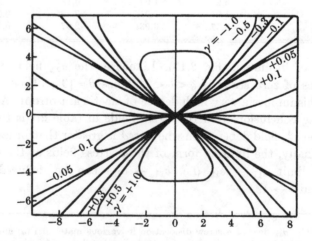

Fig. 2.14. The dilatational field of an edge dislocation in a rather anisotropic medium

## 2.7 FLEXIBLE NATURE OF DISLOCATIONS

Now let us consider a dislocation line, pinned at two points, e.g. by two impurity atoms (fig. 2.15). The dislocation line between the two points may oscillate like a stretched string. To a first approximation

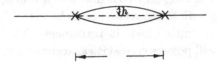

Fig. 2.15. Segment of dislocation held between pinning points (crosses) oscillating between the extreme positions shown

we may take the value of the line tension as $\sim\mu b^2$. The mass per unit length of the dislocation is $E_0/c^2$, and the speed of travel of a sound pulse along the dislocation line is

$$\sqrt{\frac{E_0}{E_0/c^2}} = c.$$

Thus this speed is equal to the speed of sound in the crystal, to the approximation that the longitudinal and transverse velocities of sound are equal, as in the Debye approximation for the specific heats. If $h$ is the amplitude of vibration, and $l$ the half-wavelength of the fundamental oscillation, one has

$$\frac{\pi^2 h^2}{4l} E_0 = kT$$

whence

$$\left(\frac{h}{b}\right)^2 = \left(\frac{4kT}{\pi^2 E_0 b}\right) \cdot \left(\frac{l}{b}\right) \sim 10^{-3}\left(\frac{l}{b}\right)$$

where $k$ is Boltzmann's constant and $T$ the temperature. Thus for a line one thousand atoms long at room temperature the amplitude of the oscillations is of the order of one atomic spacing.

To discuss more precisely the actual form of a dislocation held up at two pinning points one has to consider that (under the approximation that a dislocation has a certain energy $E_0$ per unit length, which involves treating the logarithmic factor in dislocation energy as constant), if the orientation of the straight line joining the two points is one for which $E_0$ is high, the dislocation may be able to lower its energy by bending into two segments, of greater total length but nevertheless of lower energy. The criterion for stability against this bending is most neatly expressed by using the inverse Gibbs–Wulff construction (Frank, 1963; Head, 1967) – i.e. making a polar plot of $1/E_0$ versus $\theta$. For stability in all orientations this curve must be outwardly convex everywhere. If it has concavities, then there is instability for a dislocation having the orientation of any radius between the points of tangency to a line which touches the curve at two points, and the radii to these two points have the directions of the two stable segments into which it transforms. The curve will be wholly convex, or will possess concavities, according as

$$E_0 + \frac{\mathrm{d}^2 E_0}{\mathrm{d}\theta^2} > 0$$

for all orientations $\theta$, or not. This is the condition for stability given by de Wit and Koehler (1959) but it is to be noted that if this quantity is negative in some range of orientations then the instability extends into a wider range of orientations in which this quantity becomes positive. In the case of elastic isotropy we have

$$E_0 + \frac{\mathrm{d}^2 E_0}{\mathrm{d}\theta^2} = \frac{\mu b^2}{4\pi(1-\nu)}[(1+\nu)\cos^2\theta + (1-2\nu)\sin^2\theta]\ln\frac{R}{r_0}$$

which is always positive, though (for $\nu = \frac{1}{3}$) one quarter as large for edge dislocations as for screws. With a moderate amount of elastic anisotropy (e.g. iron with $A = 2.4$) the expression $E_0 + \mathrm{d}^2 E_0/\mathrm{d}\theta^2$ becomes negative for a certain range of orientations, and straight dislocations in a certain (larger) range of orientations are correspondingly unstable.

Plainly, dislocations in the instability regions will not behave like stretched strings and may be expected to exhibit anomalous behaviour. Dynamic experiments with cyclic stress should show marked hysteresis effects. The existence of zigzag dislocation segments in anisotropic materials has been confirmed by Head, Loretto and Humble (1967) in transmission electron microscopy experiments. Elastic anisotropy also affords a very simple explanation of the observed form of faulted dislocation loops formed by electron radiation damage in the high voltage electron microscope. Brown (1967) has shown from a Wulff construction that the loops should be approximately hexagonal with edges lying along the $\langle 11\bar{2}\rangle$ directions in a (111) plane. Experiments have shown that this is indeed the case for nickel. (In copper there are additional complications concerned with dislocation dissociation – see below.)

## 2.8 ENTROPY OF A DISLOCATION LINE

As we have seen, the energy of a dislocation in typical metals is a few eV per atomic length. Cottrell (1953) discussed the question of whether a source of entropy ($\Delta S$) can be found to compensate such a large energy. Cottrell introduces the concept of the flexibility of a dislocation line: the dislocation can go from a fixed point in a lattice to one of $p$ neighbouring lattice points. If one assumes $p \sim 6$, and $n$ is the number of the planes through which the dislocation passes, one has a contribution to the free energy

$$\Delta F = T \Delta S \sim nkT \ln 6 \sim 2nkT.$$

As Cottrell remarks, this almost certainly overestimates the flexibility and should be regarded as an upper limit.

Friedel (1964) considered alternatively the contribution to the free energy due to the oscillations of the dislocation lines. If we consider that the speed of sound along the dislocation line is the same as in the crystal, the highest frequency in the spectrum of these oscillations is the maximum frequency of the Debye spectrum $\nu_m = k\Theta/h$, where $\Theta$ is the Debye temperature and $h$ is Planck's constant.

Friedel superposed this frequency and its harmonics $\nu_j = (j/n)\nu_m$ on the Debye spectrum. This overestimated the contribution to the free energy of the crystal from this source since the normal modes of the crystal, dislocated or not, are $3N$ ($N$, the number of atoms in the crystal). We would therefore argue that we must subtract $n$ modes from the Debye spectrum. The contribution to the free energy per oscillator is

$$F_{\nu_j} = \tfrac{1}{2}h\nu_j + kT \ln [1 - \exp(-h\nu_j/kT)]$$

and the total additional free energy due to the oscillations of the dislocation line, if we subtract the $n$ modes from the top of the Debye spectrum (taking them from anywhere else has less effect), is

$$\Delta F = \sum_{j=1}^{n} \left[ \frac{1}{2}\left( \frac{j}{n} - 1 \right) k\Theta + kT \ln \frac{1 - \exp(-j\Theta/nT)}{1 - \exp(-\Theta/T)} \right].$$

For $T \to 0\ °\mathrm{K}$ one has

$$\Delta F = -\frac{n-1}{4}k\Theta$$

that is, about $-\frac{1}{4}k\Theta$ per atom along the dislocation. For $T \gg \Theta$ one has

$$\Delta F \sim -nkT\left(1 - \frac{2}{3}\frac{\Theta^2}{T^2}\cdots\right)$$

or about $-kT$ per atom along the dislocation. It follows that the free energy associated with a dislocation line is positive up to the melting point, except in the case of a very large density of dislocations. When dislocations are very close together (about 10 Å apart) the factor $\ln(R/r_0)$ becomes small and the free energy can be negative at sufficiently high temperatures. In this case, the free energy decreases as the number of dislocations is increased. This may provide a formal description of melting in terms of the dislocation density increasing spontaneously in a co-operative manner. It is not very useful, however, because the whole concept of dislocations begins to lose its meaning when the dislocation density becomes very high, as we shall see when we consider the dislocation model of intercrystalline boundaries. The main point is that there is never a free energy minimum at a moderate concentration of dislocations. There is no moderate concentration corresponding to thermodynamic equilibrium as there is in the case of point defects.

## 2.9  BURGERS VECTOR

The Burgers vector of a dislocation may be defined in the following way: consider a closed circuit enclosing the dislocation line in the real crystal and let this circuit be mapped on an ideal crystal, going stepwise from one corresponding lattice point to the next one. This mapping is a locally continuable process (by 'tetrahedrulation' from

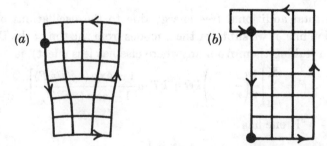

Fig. 2.16. (a) Burgers circuit in real crystal. (b) Burgers circuit mapped on to an ideal crystal

atom to atom) if, and only if, the distortion of the real crystal is not excessively large. Regions where it is possible without ambiguity are called 'good' -- other regions are called 'bad'. The circuit made continuously in good crystal may enclose a bad region and then the corresponding circuit in the ideal crystal may not be closed; the closure vector is called the Burgers vector **b** (fig. 2.16). It is necessarily a lattice translation.

In giving a sign to the Burgers vector one may adopt the so-called FS/RH convention. The direction of the dislocation line being taken as towards the observer, the circuit is made in an anticlockwise direction (thus in the sense of a right-handed screw), and the Burgers vector is the closure vector running from the finish of the circuit in the ideal crystal to its start.

It is easily verified that the Burgers vector so defined is independent of the starting point. A vector **b**′ defined by mapping **b** back into the real crystal is approximately, but not exactly, invariant. The value of **b**′ at the dislocation is to be defined, by a limiting process, as the mean of its value at surrounding points. It is sometimes, but not always, an unnecessary subtlety to draw this distinction.

For an edge dislocation the Burgers vector is perpendicular to the dislocation line. For a screw dislocation it is parallel to the dislocation line. Circuits can be quasi-continuously displaced through good regions without change of closure vector; hence one has the continuity theorem – the dislocation line cannot terminate in the interior of the crystal.

If one has two regions containing dislocations in the crystal, one finds easily that the resultant Burgers vector is the sum of the Burgers vectors of the individual dislocations (see fig. 2.17). From this it follows that junctions between dislocation lines are allowed, provided $\mathbf{b}_3 = \mathbf{b}_1 + \mathbf{b}_2$, as is seen by considering the two circuits drawn (fig. 2.18). This condition may be written $\sum \mathbf{b} = 0$, if one looks along all

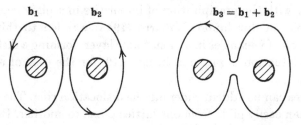

Fig. 2.17. Union of Burgers circuits to demonstrate the sum rule for Burgers vectors

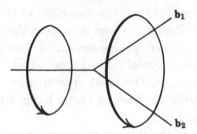

Fig. 2.18. Displacement of Burgers circuit to demonstrate the nodal sum rule for Burgers vectors

the dislocations from the junction when choosing the sense in which to take a circuit.

## 2.10 MOTION OF DISLOCATIONS

A straight edge dislocation has a rigorously defined plane in which it can move conservatively, without the addition or removal of atoms. This plane, called the slip plane, includes **b** and the dislocation line. Likewise, polygonal or curved edge dislocations have rigorously defined slip surfaces and slip on these surfaces is called prismatic slip. Non-conservative motion, or climb, as it is called, can only occur in the presence of point defects. In general, a slip plane is not defined for screw dislocations since **b** and the dislocation line are co-linear. However, this is not the case in all crystal structures, owing to dissociation (see section 2.13).

## 2.11 INTERACTION BETWEEN DISLOCATIONS AND POINT DEFECTS

In analogy with the precipitation of impurities in a plane, occurring, for example, in duralumin, Nabarro (1948) was led to think of a condensation of vacancies in a monatomic layer, forming a dislocation ring. This situation is pictured in fig. 2.19, for the case of a simple cubic lattice.

Vacancies can be added, widening the dislocation ring. The dislocation ring can easily glide from one lattice plane to another. However, no metal crystallises in the simple cubic structure; for this reason,

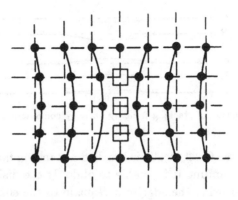

Fig. 2.19. Condensation of vacancies (squares) to form a dislocation loop

let us study the situation in the f.c.c. lattice. The close-packed planes in this lattice are the (111) planes: their stacking is of the type △ △ △ as illustrated in fig. 2.20.

The two permissible modes of stacking designated by symbols △

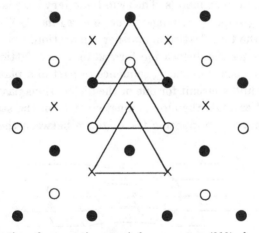

Fig. 2.20. Projection of consecutive atomic layers on to a (111) plane. ● first layer, ○ second layer, × third layer

and ▽ lead to three possible projected positions of the close-packed planes, designated $A$, $B$, $C$. The cyclic order $ABC$ corresponds to △ stacking, cyclic order $CBA$ to ▽ stacking. Hexagonal close packing corresponds to △▽ stacking in alternation, or $ABAB$. Let us suppose vacancies condense in an $A$ plane: the planes above and below this plane will collapse, and we shall have a plane $B$ above in contact with a plane $C$ below, namely a ▽ stacking interposed in the

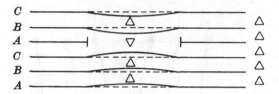

Fig. 2.21. Formation of stacking fault by vacancy condensation in an f.c.c. lattice

$\triangle$ sequence (see fig. 2.21). This is called a stacking fault. The dislocation ring surrounding it is unable to glide. If the dislocation were to
glide it would bring the edge of a $B$ plane to the edge of an $A$ plane
and since $B$ and $A$ are different planes of atoms they will not fit.
This glide would create a surface of high energy, of the same order of
magnitude as the energy of a grain boundary in a typical polycrystalline metal. The stacking fault itself has finite energy, of the order ten
times smaller than the grain boundary energy, as we infer from the
measurable energy of twin boundaries, having a closely related
configuration, in f.c.c. metals. The twin boundary is a place where the
regular $\triangle \triangle \triangle$ sequence changes over to a regular $\triangledown \triangledown \triangledown$ sequence,
representing the f.c.c. lattice in another orientation.

Let us consider the dislocation formed in a f.c.c. lattice when a $\triangledown$
stacking is interposed in the $\triangle$ sequence in part of a plane. Now it is
impossible to find a circuit for one of the dislocations passing entirely
through good crystal: the circuit must start from the stacking fault
as shown (fig. 2.22). Supposing the distance between the $\triangledown$ stacked

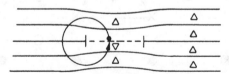

Fig. 2.22. Burgers circuit (the circle starting and ending on the fault) round one of the
dislocations terminating a stacking fault

planes to be equal to the ordinary distance between (111) planes,
one finds as a Burgers vector $-\frac{1}{3}[111]$L and $\frac{1}{3}[111]$R for the left and
right sides of the diagram. The values $\frac{1}{3}[111]$L and $-\frac{1}{3}[111]$R are not
allowed when there is only one $\triangledown$ fault, but can occur for a double
fault, which might be formed by an aggregation of interstitial atoms
(see fig. 2.23). Dislocations of this kind, around which a Burgers

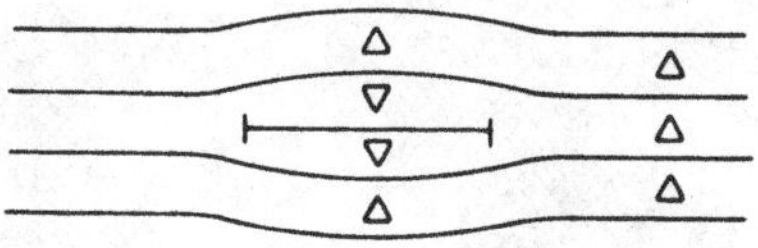

Fig. 2.23. Faulted dislocation loop formed from interstitials

circuit cannot be completed in good crystal, being interrupted by one or more fault surfaces, and for which the resultant Burgers vector is accordingly not in general a lattice vector, are called 'imperfect' or 'partial' dislocations.

Owing to their dilatational fields point defects diffuse to both screw and edge dislocations. The arrival of point defects at an edge dislocation causes the formation of small steps or jogs (section 2.27) by up- or down-climb. Point defects are believed to diffuse particularly easily along dislocations: this process is called pipe diffusion.

## 2.12 INTERACTION OF PARALLEL DISLOCATIONS

During the processes of slip or climb dislocations move into low energy configurations under the influence of their mutual attraction or repulsion. The nature of these processes depends critically on the velocity of motion at very high velocities, but at moderate or slow speeds we can deduce the nature of the interaction from the static elastic fields of the interacting dislocations. During plastic deformation one of the most common processes is the interaction of like or unlike dislocations. Consider the interaction of parallel edge dislocations through the strain field of fig. 2.10. Unlike dislocations form stable configurations (so-called dipoles) at 45° to the slip plane. Trains of edges of opposite sign interleave to form large numbers of dipoles (multipoles). Like edge dislocations on different slip planes form stable dislocation walls (polygonisation) perpendicular to the slip planes. There is ample experimental evidence for these low energy forms (fig. 2.24). At high temperatures (relative to the melting point), when the edge dislocations can climb, the edge dipoles and multipoles vanish, while the polygonised walls, which are stable under these conditions, sharpen and become more regular. Anisotropic elasticity does not generally affect these conclusions, although the equilibrium angles of the stable configurations are changed. In a few cases of

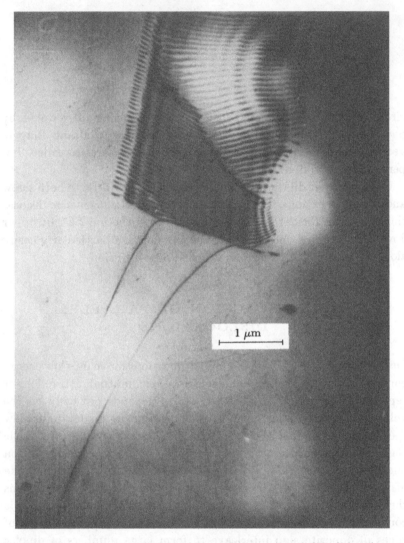

Fig. 2.24. Edge multipole group formed by mutual trapping of glide dislocations on close slip planes. (Steeds and Hazzeldine, 1964)

very high anisotropy some of the stable edge configurations can vanish (in zinc for example). However, screw dislocations which, according to isotropic elasticity, form no stable low energy configurations, can do so in some cases when anisotropic elasticity is considered (see section 2.6) so long as the screws are constrained to move in parallel slip planes.

Another conclusion changed by taking anisotropic elasticity into account is that concerning the interaction of parallel edge and screw dislocations. According to isotropic elasticity, since the dislocations have perpendicular Burgers vectors there will be no interaction. The interaction does not necessarily vanish in the anisotropic case, and consideration of the action and reaction of the two dislocations leads to equations (2.6).

## 2.13 EXTENDED DISLOCATIONS

When part of a dislocation line in a f.c.c. crystal having Burgers vector $\frac{1}{2}[1\bar{1}0]$ lies in a (111) plane, that part may dissociate into two glissile half-dislocations with Burgers vector $\frac{1}{6}[1\bar{2}1]$L and $\frac{1}{6}[2\bar{1}\bar{1}]$R. These dislocations repel each other and, as they separate, a sheet of stacking fault is formed in the slip plane between them (see figs. 2.25 and 2.26). The energy of this fault prevents them from separating too

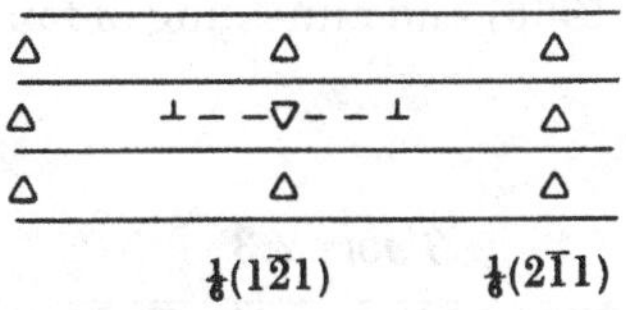

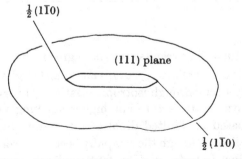

Fig. 2.25. Stacking sequence through a dissociated dislocation in an f.c.c. crystal

far. This gives a constant attraction, balanced against the repulsion, proportional to the scalar product of the Burgers vectors of the two partial dislocations, and inversely proportional to the distance

$\frac{1}{2}(1\bar{1}0)$

(111) plane

$\frac{1}{2}(1\bar{1}0)$

Fig. 2.26. Dissociation of a perfect dislocation in the (111) plane of an f.c.c. crystal

between them. Equilibrium widths of 10 or 15 atomic spacings are estimated for copper, but of only about 2 atomic spacings for aluminium, which has a higher stacking fault energy. The stacking fault illustrated in fig. 2.25 is called intrinsic. Faults of the form … △▽▽△△△ … are also encountered and are called extrinsic faults. They are less common than intrinsic faults, probably because of a high core energy for the partial dislocations in the extrinsic case. Clarebrough and Head (1969) have demonstrated that the partial dislocations may have instability regions, while the perfect dislocation formed by constriction of the fault is stable. (Examples of dissociated dislocations are shown in chapter 3.)

Many different methods of measuring stacking fault energy have been devised by direct observation in transmission electron microscopy. These involve the observation of the equilibrium shape of extended nodes (Whelan, 1959b; Howie and Swann, 1961; Brown, 1964), stacking fault tetrahedra (Seeger, 1964; Jøssang and Hirth, 1966; Clarebrough, Humble and Loretto, 1967), faulted dipoles (Steeds, 1967; Häussermann and Wilkens, 1966), intrinsic/extrinsic fault pairs (Gallagher, 1966) and annealing of faulted loops (Edington

*Table* 2.3

| Metal | Stacking fault energy ($ergs\,cm^{-2}$) |
|:-----:|:-----:|
| Ag | 22 |
| Cu | 55 |
| Au | 50 |
| Al | 200 |
| Ni | 250 |
| Pb | 30 |

and Smallman, 1965). The results are in fair agreement for low stacking fault energy metals and alloys ($\lesssim 20$ ergs $cm^{-2}$) although the interpretation has mostly been through isotropic elasticity theory which may have led to a systematic error. For higher stacking fault energies only methods based on faulted dipoles or the annealing of faulted loops can be applied, although these rapidly become uncertain through the failure of linear elasticity. The faulted dipole method has the advantage of simple treatment using anisotropic elasticity. A few results are shown in table 2.3; a very thorough review of the experimental situation was recently made by Gallagher (1970). Since then, however, the weak beam technique has been developed and used to obtain more accurate values of ribbon-widths (see section 3.5.2).

## 2.14 CROSS-SLIP

The glide of a dissociated screw dislocation is confined to the plane in which it has extended to form a ribbon of stacking fault; however, by constriction, i.e. bringing the pair of partial dislocations together and reducing the width of the stacking fault ribbon to zero, it is enabled to extend, and then glide, in another plane. This process is known as cross-slip. (For example of cross-slip see fig. 3.16.) Energy reduction processes for screw dislocations during slip often depend on cross-slip, which is made more difficult as the stacking fault energy decreases. Consequently stacking fault energy has a marked influence on plastic properties, and the dislocation arrangement after deformation is much simpler in low than in high stacking fault energy metals and alloys. It has further been suggested that impurities may sometimes segregate to stacking faults, lowering their energy and changing the mechanical properties.

## 2.15 ADDING OF DISLOCATIONS

A pair of dislocations will add if the new dislocation has smaller energy than the original pair. From this principle, one finds that, if the isotropic elasticity theory is applied, and if the core energy and the difference between screw and edge are neglected, dislocations will add if their Burgers vectors make an angle smaller than 90°.

## 2.16 DISLOCATION LOCKS†

Let us consider two extended dislocations, $\frac{1}{2}[10\bar{1}]$ and $\frac{1}{2}[011]$, gliding on different intersecting planes (see fig. 2.27(a)). The partials $\frac{1}{6}[2\bar{1}\bar{1}]$ and $\frac{1}{6}[\bar{1}21]$ will interact at the junction to form a partial $\frac{1}{6}[110]$. This partial repels both the partials $\frac{1}{6}[11\bar{2}]$ and $\frac{1}{6}[112]$. Thus a bent ribbon of extended dislocations, which is unable to glide in any

† See also section 3.4.1.

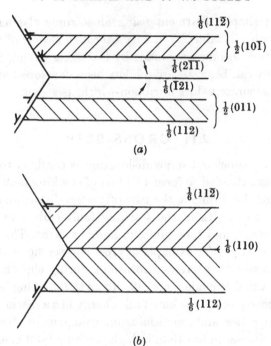

Fig. 2.27. Combination of dissociated dislocations on intersecting glide planes to form a Lomer–Cottrell lock

direction, has formed (see fig. 2.27($b$)). For an alternative but similar reaction we consider the interaction of the dislocations $\frac{1}{2}[10\bar{1}]$ and $\frac{1}{2}[\bar{1}0\bar{1}]$. The partials $\frac{1}{6}[\bar{2}11]$ and $\frac{1}{6}[2\bar{1}1]$ will in this case combine to give the partial ('stair-rod') $\frac{1}{3}[001]$. The energy reduction in this case (the Hirth lock) is less than the former case (the Lomer–Cottrell lock, of which some examples are shown in fig. 3.21.) Experimentally, very few Hirth locks are observed but Lomer–Cottrell locks are quite common. However, their stability depends to some extent on the stacking fault energy, widely dissociated locks being very stable. In copper, where the stacking fault energy is quite high, stress concentration can cause constriction of the Lomer–Cottrell lock and glide on {001}. Faulted dipoles are also unable to move (sessile) and may be important locking configurations.

The importance of these locks lies in their ability, under favourable circumstances, to cause pile-ups of dislocations (fig. 3.19) which contribute to work hardening. The head of a screw pile-up is a particularly favourable place for cross-slip to occur.

## 2.17  INTERCRYSTALLINE BOUNDARIES

Figure 2.28 shows the simplest type of tilt boundary in which two crystals of simple cubic lattice, slightly misaligned to each other, are joined together. Such a boundary involves a large number of parallel edge dislocations. This configuration is stable in respect of glide motions, as can be easily seen from the distribution of shear stress around an edge dislocation, described in section 2.4.

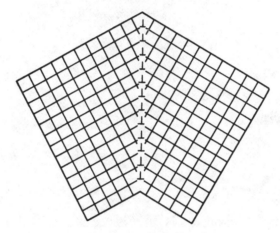

Fig. 2.28. A simple tilt boundary

So long as the angle of misalignment $\alpha$ is small, so that the spacing $d$ between the dislocations (given by $b/d = \alpha$) is large, the dislocations are distinct individual dislocations, They have been observed by transmission electron microscopy (fig. 2.29; see also fig. 3.16). The dependence of boundary energy on misalignment may be calculated from the dislocation model, in satisfactory agreement with observation; the observed values may then be used to estimate the core energy of the dislocations, as mentioned earlier. The agreement is actually satisfactory up to higher dislocation densities than we have a right to expect.

In the case of 'twist boundaries', in which the misalignment is by rotation about an axis normal to the boundary plane, the corresponding dislocation system is a crossed grid of screw dislocations: or, at surfaces of trigonal symmetry, an equivalent hexagonal network made up of screw dislocations of three types.

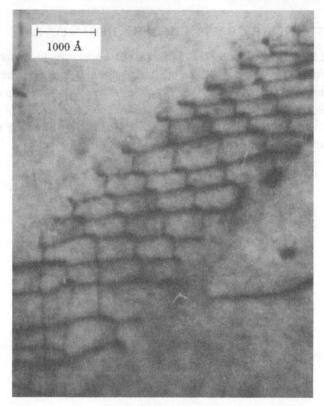

Fig. 2.29. Dislocation network constituting a low angle boundary in pure iron.
(By courtesy of D. J. Dingley)

In the general case of misalignment by a rotation $\alpha$ about an axis defined by the unit vector $l$, the dislocation content of the boundary is such that if $r$ is the vector connecting two distinct points in the boundary (which is not necessarily plane), any line in the boundary connecting these two points crosses dislocation lines having the same sum of Burgers vectors $\sum b$, and

$$\sum b = r \times 2l \sin \tfrac{1}{2}\alpha.$$

The proof of this formula is easily given by constructing a Burgers circuit intersecting the boundary at the two specified points. So long as the misalignment is small, and the dislocation density accordingly low, this may be done in such a way that the circuit passes continuously through 'good' crystal. This is no longer so when the misalignment is large, and alternative specifications of the misalignment and dislocation content then arise. A cubic crystal may be rotated into parallelism with another in 24 different ways. If we always choose the rotation of smallest angle, the largest value the angle can take is about 63° and then there are four alternative equal rotations. As a simple extreme example one may consider the symmetrical tilt boundary about a cube axis when the angle $\alpha$ goes to 90°. This can either be described as having an atomic density of dislocations or no dislocations at all (see fig. 2.30).

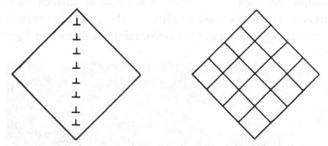

Fig. 2.30. The formal limit of a tilt boundary with 90° tilt is equivalent to many dislocations or none at all

It is this kind of ambiguity which destroys the usefulness of describing a liquid as a crystal full of dislocations – the very high density of dislocations has no unique specification.

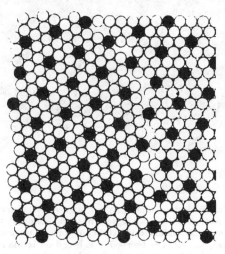

Fig. 2.31. Coincidence lattice (full circles) at a grain boundary

One specification of large angle grain boundaries is the coincidence model due to Kronberg and Wilson (1949). This model proposes that grain boundaries where the atoms in either grain have a large number of coincidences with the lattice of the other have particularly low energy and are most mobile. An example of a coincidence boundary separating two grains misoriented 38° or 22° about ⟨111⟩ is shown in fig. 2.31. One in seven of the atoms of one grain coincides with the lattice of the other. When the grains depart from the coincidence configuration the extra distortion results from dislocations which may conveniently be defined relative to the coincidence lattice. An example of some grain boundary dislocations is shown in fig. 2.32.

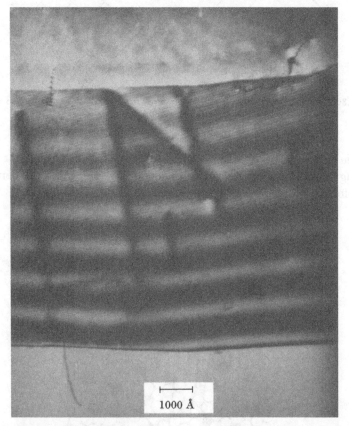

1000 Å

Fig. 2.32. Transmission electron micrograph of a grain boundary in Al (by courtesy of R. Pond). Isolated dislocations can be seen near the boundary (narrow images) but when they run into the boundary quite different (broad) images are observed. Two sets of parallel fringes can also be observed. The broad set is due to depth oscillations, while the narrow set arises from grain boundary dislocations

## 2.18 FORCES ON DISLOCATIONS

Consider a dislocation moving in a crystal and passing between two initially adjacent atoms, $P$, $Q$, so as to enter a Burgers circuit constructed through these points. As it does so, the closure vector of the mapped circuit in the ideal crystal changes by $\mathbf{b}$. But its only alteration is between the map points of $P$ and $Q$. These points have therefore suffered a relative displacement $\mathbf{b}'$. The motions we consider in this argument need not be restricted to motions in the glide plane. The stress in a body is measured by the stress tensor $\mathbf{P} = P_{ij}$, so defined that $|\mathbf{da}|$ times the traction on an element of area $\mathbf{da}$ (i.e. the force required to be applied at the faces of this element of area to prevent displacement if a cut is made at it) is $\mathbf{P}\,\mathbf{da} = \sum_j P_{ij}\,\mathrm{d}a_j$.

Then the work done by the elastic field when a dislocation line element $\mathbf{ds}$ sweeps the element of area $\mathbf{da}$ is

$$(\mathbf{P}\,\mathrm{d}a)\cdot\mathbf{b}' = \sum_i \sum_j P_{ij}\,\mathrm{d}a_i\,b'_j = (\mathbf{P}\mathbf{b}')\cdot\mathrm{d}a.$$

Dividing by $|\mathbf{da}|$ gives the force due to the elastic stress field tending to move the dislocation line element $\mathbf{ds}$ in that direction in which it sweeps the area $\mathbf{da}$. The force per unit length may be seen either as the scalar product of the traction on the swept area with the Burgers vector, or $|\mathbf{b}|$ times the scalar product of traction on a plane normal to the Burgers vector with the unit vector normal to the swept area. Considering all the planes in which the line element $\mathbf{ds}$ can move we see that this corresponds to the existence of a force $(\mathbf{P}\mathbf{b}') \times \mathbf{ds}$ on each element $\mathbf{ds}$ of the dislocation line.

## 2.19 THE FORCE FOR GLIDE MOTION

The glide motion of a dislocation is one causing no change of volume of the crystal; commonly the glide surface is a plane. The tangential component of traction on a plane is called the shear stress on that plane. The shear stress on a glide plane with unit normal $\mathbf{n}$ is $\sigma_j = \sum_i P_{ij} n_i$, and we have from the foregoing formula that $\boldsymbol{\sigma}\cdot\mathbf{b}$ measures the force per unit length on the dislocation line tending to move it in its glide plane. This is a uniform force normal to the dislocation line, independent of dislocation orientation. In conjunction with the line tension approximation it gives us a close analogy, in two dimensions instead of three, between the equilibria of dislocations under stress and soap bubbles under pressure.

## 2.20 FORCES FOR CLIMB MOTION

The force $(\mathbf{Pb}') \times d\mathbf{s}$ also contains in general a component normal to the glide plane of the dislocation. It is readily visualised as the force tending to 'squeeze out' the inserted half-plane. Such motion, however, involves a volume change, which can only be brought about by creation or absorption of interstitial atoms or vacancies at the dislocation. Such motions require diffusion processes, are relatively slow, and do not occur at low temperature. If the concentration of point defects is not in equilibrium, this gives another force, the osmotic force, on the dislocation, in addition to that due to elastic stresses. The osmotic force due to a concentration of vacancies which exceeds the equilibrium concentration by a factor $\alpha$ is

$$(\mathbf{b}' \times d\mathbf{s}/v')\,kT\ln\alpha$$

where $v'$ is the lattice volume per atom.

We now discuss briefly a number of topics which cannot be dealt with in detail here but are too interesting to leave unmentioned.

## 2.21 PILE-UP EQUILIBRIUM

In a slip band we have a number of dislocations in the same or nearly the same glide plane, each acted upon by the same force from the applied stress, and repelling each other with a force inversely proportional to distance. If the leading dislocation is held up by a barrier of some sort, we have the problem of finding their equilibrium distribution. This looks like being a very difficult problem for a large number of dislocations, but is actually very simple: the solution was given by Stieltjes in 1885 (without reference to dislocations, of course). The theory was applied to dislocations by Eshelby, Frank and Nabarro (1951) and has since been widely used for different sorts of pile-ups, subject to various constraints and arranged in a number of regular arrays (see the book by Nabarro (1967) for a review, and more recent work by Smith (1971)). Arrays with smaller numbers of dislocations where the stress fields of the individual constituents are important have been investigated by Hazzledine and Hirsch (1967). The problem has received a lot of attention due to its connection with the work hardening process (see chapter 5; an example of a pile-up at a grain boundary is shown in fig. 3.19).

## 2.22  THE ORIGIN OF DISLOCATIONS

This is something we do not know enough about. There are probably mechanisms of dislocation production associated with inhomogeneous distribution of impurities similar to those deduced by Ashby and Johnson (1969) in their analysis of results from a study of $SiO_2$ in copper. For pure materials the principal process preventing us from obtaining large dislocation-free specimens is probably the formation of dislocations by vacancy aggregation, both in the moving temperature gradient while the crystal is being grown and during the process of cooling down after growing or annealing. For reasonable cooling times it is only in very thin specimens (especially metal whiskers) that there is time for vacancies to diffuse to the free surface. The interesting and difficult problem is to decide whether they should then aggregate into dislocation loops or open cavities. The problem is difficult because the fate of the aggregate is probably decided when it consists of approximately 10 vacancies – too small a number to be considered macroscopically and a large number to be considered discretely. Some discussion of the matter will be found in a paper by Fisher and the discussion following it in the report of the 1956 Lake Placid Conference, and more recently discrete calculations have been attempted by Johnson (1967).

## 2.23  FORMATION OF KINK BANDS

When a pair of boundaries of opposite sign are close together there must exist a shear stress of the form shown in fig. 2.33 to keep the rows of dislocations apart. The region $A$ can be stressed sufficiently to create new dislocation pairs extending the kink band.

Shear stress

Fig. 2.33. Dislocations and stress concentration region $A$ at the edge of a kink band (a similar situation exists at the edge of a deformation twin or a martensite plate)

The theory of this process, which is important in twinning and martensitic transformations, has been discussed by Frank and Stroh (1952), and in the book of Friedel (1964).

## 2.24  ADSORPTION AND PRECIPITATION AT DISLOCATIONS

The first application of this idea was by Cottrell (1948) who explained various phenomena in mild steel by the adsorption of carbon atoms at dislocations. We now have many examples of the precipitation of impurities at dislocations, making the dislocations visible. This is attributable to three concurrent effects: the adsorption of impurities at dislocations, the enhanced diffusivity along dislocation lines, and reduced strain energy of nucleation in an already severely strained lattice.

## 2.25  INTERACTION OF ELECTRONS WITH DISLOCATIONS

This is a topic deserving much fuller treatment but we shall mention only two basic ideas of the subject. The publications of Read (1954a, b, 1955) deal specifically with dislocations in covalently bonded crystals in which we suppose there are what Shockley (1953) calls 'dangling bonds' along the dislocation. These are regarded as essentially equivalent to impurity centres, with some important distinctions arising from their relatively dense distribution along a line. Kawamura (1958) has suggested that in a more general class of crystals, e.g. KCl, the deformation potential may suffice to give line-bound states for electrons.

## 2.26  THE WORK HARDENING CURVE

This was the subject of Taylor's original paper in 1934 and, together with metal creep, has been the subject of more studies in dislocation theory than anything else; however, the full truth of this complex matter has yet to be said. A discussion is presented in chapter 5.

## 2.27  JOGS IN DISLOCATIONS

A place on the dislocation line where it passes from one glide plane to an adjacent parallel one is called a jog. If the dislocation is nonscrew, then we can imagine that it was formed by insertion of a layer of atoms into a cut in the crystal. The edge of this layer is the dislocation line, and is stepped wherever there is a jog in the dislocation. To describe a given dislocation, the choice of cut and inserted layer is not unique, except as to the position of its edge.

As an example let us consider an ionic crystal with the NaCl structure. Three equivalent choices for the inserted layer to make a dislocation having Burgers vector $\mathbf{b} = (a/2)[110]$ are shown in fig. 2.34.

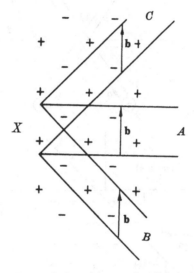

Fig. 2.34. Alternative 'inserted planes' for an edge dislocation in an ionic crystal

The layers shown, chosen from many other possibilities, contain two extra (110) planes in $A$, or one extra (100) or (010) plane at $B$ or $C$. In each case the dislocation lies at the edge of the inserted layer, along [001] at $X$ in fig. 2.34. It is an edge dislocation for the slip plane $(1\bar{1}0)$. In each case the faces of the cut must be separated along the same direction and by the same amount along this direction, $\mathbf{b} = (a/2)$ [110], and in each case the contents of the layer have the same projection on to the plane normal to $\mathbf{b}$; see fig. 2.35(a). Encircled ions in this figure are at an odd, the remainder at an even number of (110) interplanar spacings above the projection plane. The dislocation lies along $AB$. Suppose now that this dislocation is intersected by another dislocation whose Burgers vector $\mathbf{b}_2$ may have any of twelve

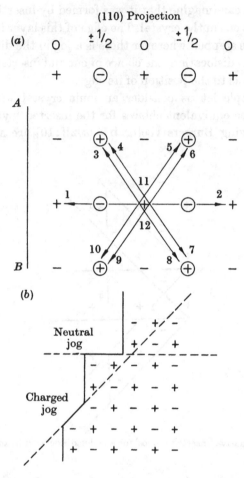

Fig. 2.35. Various possibilities for the formation of jogs by the intersection of dislocations in an ionic crystal (see text)

different orientations represented by the ten arrows on fig. 2.35(a),
together with two directions perpendicular to the plane of the paper.

The intersecting dislocation makes a relative displacement $b_2$ of the
two portions of crystal either side of its glide plane. As it does this to
the inserted layer also, it generally makes a jog in the first dislocation.
Except at the edge, the place where the glide plane cuts the inserted
layer depends on where we imagine this inserted layer to be. It is
advantageous to imagine it as one of the single planes (100) or (010),
so that the intersection is a straight line.

None of the possibilities for $b_2$ is parallel to the line of the first dislocation, so that a jog is produced in every case. Numbers (11) and (12), with $b_2$ parallel to $b_1$, make a 'slip-jog', which can disappear again by glide. Numbers (1) and (2) make neutral jogs, with both a positive and a negative ion at the step (fig. 2.35($b$)). The remaining eight cases, numbers (3) to (10), make a charged jog together with a jog component which can glide out again. The charged jog carries an effective charge of $+\frac{1}{2}e$ or $-\frac{1}{2}e$ according to whether the ion at the step is positive or negative. Note that the conclusion differs from that of Seeger (1955$b$) who failed to consider all the possible types of intersection.

Experimental evidence bearing on the presence of charged jogs may be found in experiments by Stepanow (1933) and by Fishbach and Nowick (1958), showing charge transfers accompanying plastic deformation in NaCl, in addition to the temporarily enhanced conductivity discovered by Gyulai and Hartley (1928).

In addition to the intersectior of dislocations, jogs are commonly formed by the local cross-slip of screw segments during plastic deformation. The resulting jogs are typically 100$b$ in length and are sometimes called super-jogs.

## 2.28  JOG DRAGGING

Consider a closed circuit into which a right-handed screw dislocation is moved. The circuit is then opened as shown in fig. 2.36, producing a clockwise simple shear.

Consider now the intersection of two screw dislocations. Two intersecting right screws produce overlapping material (a line of interstitials). The same applies for two left screws – a mirror image of the previous case (fig. 2.37($a$)). If one screw is right and the other left a line of vacancies is formed (fig. 2.37($b$)). Cottrell (1957) has pointed out that if the intersecting dislocations are activated by the same stress there is a greater probability of the interstitial-forming mechanism occurring.

The case of super-jogs is less simple since dissociation of the jog will occur leading to complicated configurations, some of which are sessile and some glissile under the action of an applied stress. Hirsch (1962) has discussed the problem on the assumption of intrinsic stacking faults and Weertman (1963) has extended the analysis to the case of extrinsic faults.

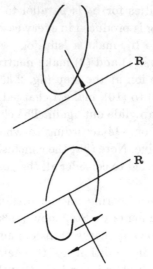

Fig. 2.36. Entry of a dislocation into a Burgers circuit

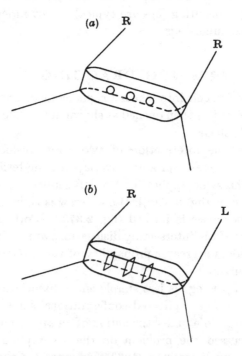

Fig. 2.37. Formation of (a) interstitials (circles) or (b) vacancies (squares) by intersection of screw dislocations

One must, however, enquire whether these processes occur at all.

Consider an expanding dislocation loop acquiring jogs by intersection with other dislocations. Each jog is a little piece of dislocation

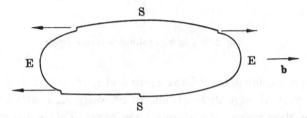

Fig. 2.38. Jogs formed on a glide loop. Edge (E) and screw (S) parts of the loop are marked

which can glide in the directions shown in fig. 2.38. Let us consider the force which makes the jog gli.e. Apart from a component directly due to the applied stress acting on the jog, which is equally likely to assist or oppose its motion, this force must come from the line tension of the two portions of dislocation line to right and left of the jog, making an angle with each other (see fig. 2.39).

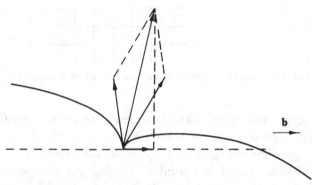

Fig. 2.39. The resultant force on a dislocation jog and its resolved component in its direction of motion parallel to the Burgers vector

Now we ask if there is any stationary shape in which the dislocation line can move, dragging the jog along with it. We conclude that there cannot be, because this requires the dislocation line to be moving faster at a place where it has convex curvature than at a place where it is relatively straight. This conclusion must be wrong, and when the main part of the dislocation line continues for a while in steady

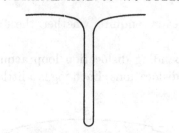

Fig. 2.40. Jog lags behind to create dipole

motion its jog must hang back more and more (fig. 2.40). This produces a pair of edge dislocations, of opposite sign, attracting each
other, in glide planes one atomic layer apart. This is just equivalent
to a line of vacancies (fig. 2.41), or to a line of interstitial atoms for a
jog of opposite sign. If the temperature is not too low, such a line will
be expected to evaporate into independent point defects.

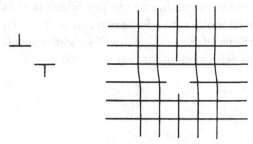

Fig. 2.41. Equivalence of a narrow dipole to a line of vacancies

Other suggestions which have been made to account for Gyulai's
temporary enhancement of conductivity depend on producing
the corresponding configuration by accidental meeting of a pair
of edge dislocations of independent origin, one (or perhaps two
or three) atom planes apart; or, alternatively, they depend on the
localised energy release when such a pair meet in the same plane and
annihilate each other. In the case of super-jogs on screw dislocations
the jog dissociation leads to sessile configurations which pull out
edge dislocation dipoles as the screw glides and which are stable
except at high temperatures. On relaxing the applied stress the
dislocation slips back along the glide plane but is retarded by previously glissile jogs becoming sessile. The resulting dislocation configuration is a common experimental observation (fig. 2.42).

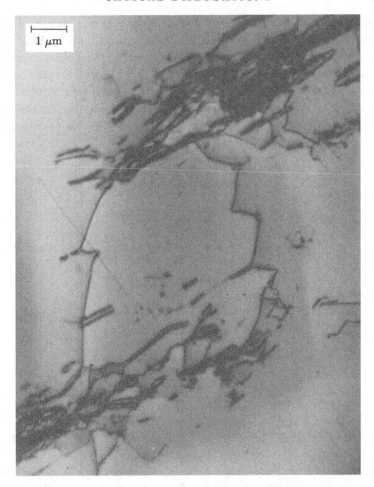

Fig. 2.42. Transmission electron micrograph of dislocations in a (111) section of a fatigued copper single crystal. Two long dislocations can be seen which have predominantly screw character. These dislocations are heavily jogged. (By courtesy of C. A. McCombie)

## 2.29 SPIRAL PROCESSES

Under this term we will group together several different processes which are geometrically related to each other. The first is crystal growth. The Volmer theory of crystal growth, which is the right theory for undislocated crystals, does not fit the usual experimental facts. This theory assumes independent nucleation of new monolayers on the crystal surface and spreading of these layers by addition of atoms along the step which is the edge of the layer. It leads to a nucleation rate which is very sensitive to supersaturation. In any crystal growing from a solution, the supersaturation is highest at the corners because of diffusion. New layers ought therefore to nucleate at the corners before there is time for them to spread across the face. Crystals growing by this mechanism will therefore normally be dendritic, and will require for growth a higher supersaturation than is usually observed to be necessary. The reason is that crystals

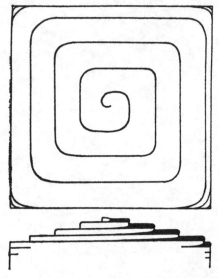

Fig. 2.43. Spiral growth terrace on a crystal containing a screw dislocation

containing screw dislocations have permanent steps in the surface. They do not require nucleation of new layers, because the crystal consists of one helicoidal layer only. Atoms continuously add themselves to the step, making it rotate around its fixed point at the end of the dislocation. In doing so the step does not remain straight, like the hand of a clock, because there is no reason why its speed should increase in proportion to its distance from the centre. Its speed is much more nearly constant along its length, and it therefore winds itself up into a spiral (fig. 2.43). Near the centre of the spiral it acquires the same curvature as the critical nucleus of Volmer's theory, which is in unstable equilibrium with the supersaturation.

Suppose we now have a right and left screw emerging close together at the surface. If they are too close together for the equilibrium curvature (so that the Volmer critical nucleus can not pass between them) no growth occurs. The curved step is in stable equilibrium with the supersaturation. Increased supersaturation increases the curvature, and beyond a critical value growth proceeds. Spirals commence to form around both dislocations ends, but two steps meeting face to face form a level surface. Then we have a closed step

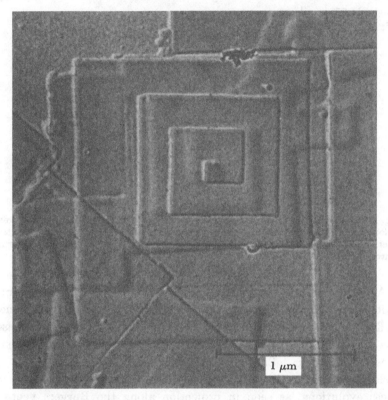

Fig. 2.44. Shadowed replica of a growth spiral formed in polyethylene oxide. (Lotz *et al.*, 1966)

ring around the dislocation pair, and a short length of step, which can repeat the process, connecting them. This provides the basis of a successful quantitative theory of crystal growth rates. The corresponding surface configurations have now been observed on a great variety of crystals (fig. 2.44).

Exactly the same geometry applies to certain kinds of dislocation motion in the interior of crystals. Under shear stress a dislocation line held at two points bows out in its glide plane like a two-dimen-

sional soap-bubble. Increased stress brings it to a maximum curvature after which it can expand continuously with decreasing
curvature, repeating the configuration of fig. 2.45. This (the so-called
Frank–Read source mechanism) is believed to be one of the most
important processes by which dislocations are multiplied in plastic
deformation. The first unambiguous observation of it was made by
Dash (1957) in silicon.

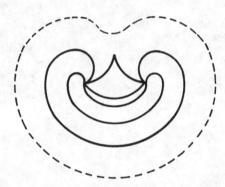

Fig. 2.45. Generation of closed loops of growth step from a pair of opposite screw
dislocations. The same figure is applicable to generation of dislocation loops in the
interior of a crystal, in a slip plane (Frank–Read source) or in climb (Bardeen–Herring
source)

A dislocation line can similarly be bowed out by the osmotic force
in its climb plane and perform the evolutions of fig. 2.45 in this plane,
as first suggested by Bardeen and Herring (1952). However, this case
has an interesting variant the idea of which is due to the Ghent
group, Amelinckx, Bontinck, Dekeyser, and to Seitz (1957). A
dislocation intermediate between edge and screw should undergo
these evolutions, as seen in projection along the Burgers vector,
except that where they meet and cancel each other in the other
cases, they are now actually far apart and pass each other by. The
process can then develop a helix of many turns. This is one way in
which the helical dislocations found by the Ghent workers can be
explained. We think there is perhaps another explanation, chiefly
because their helices, first observed in $CaF_2$, with evaporated silver
on the surface, heated in hydrogen, are so accurately oriented along
screw dislocation directions.

A pure screw dislocation is not able to climb, but if on account of
thermal oscillations a segment deviates from the straight dislocation
line, segments with edge components appear and climb can occur.

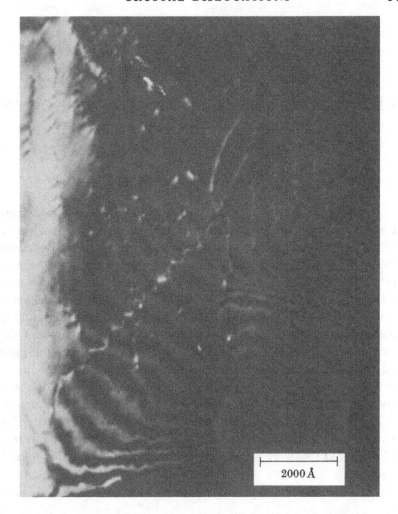

Fig. 2.46. Dark-field transmission electron micrograph showing a helical dislocation produced in a Cu–10at.%Al alloy by point defects generated during bombardment by 750 keV electrons

Being of opposite sign, these segments will climb in opposite directions. Thus the straight screw dislocation can be transformed into a helix, a process which has been observed during the process of electron radiation damage of metals in the high voltage electron microscope (see fig. 2.46). Examples in quenched metals are shown in figs. 3.11 and 3.31.

CHAPTER 3

# OBSERVATIONS OF DEFECTS IN METALS BY ELECTRON MICROSCOPY

*by* M. J. WHELAN†

## 3.1 INTRODUCTION

It is now just over ten years‡ since the first observations of dislocations in metal foils by transmission electron microscopy were reported by Hirsch, Horne and Whelan (1956) and independently by Bollmann (1956). Since that time the technique has been applied to such a variety of problems in crystal and metal physics and the literature on the subject is now so voluminous that it is impossible in a chapter of this length to give an exhaustive review. Instead the author intends to concentrate first on an exposition of the principles of the technique (sections 3.2 and 3.3) both from the practical point of view and from the point of view of theories of interpretation of image contrast. Some applications to various problems in metal physics will then be described (sections 3.4 and 3.5), which will mainly be illustrative of work done in the Cavendish Laboratory. For this the author makes no apologies in view of the above remark, believing that this work is fairly typical of current applications of the technique.

## 3.2 EXPERIMENTAL TECHNIQUES

### 3.2.1 Preparation of thin specimens for transmission electron microscopy

Metallic specimens suitable for examination by transmission of electrons in the electron microscope must be of the order of a few (about two) thousand Ångström units in thickness or less depending

---

† Dr Whelan is Reader in the Physical Examination of Materials at the Department of Metallurgy and Science of Materials, University of Oxford.
‡ This contribution was prepared in 1967; some minor up-dating was made and references added in 1973.

on atomic number. Several methods exist for preparing thin specimens:

1. Chemical deposition and electrodeposition from solution.
2. Vacuum evaporation onto a suitable substrate followed by
   stripping.
3. Thinning of molten liquid droplets by the surface tension effect.
4. Ion bombardment etching.
5. Cutting thin sections by microtome.
6. Mechanical beating.
7. Etching and chemical polishing.
8. Electropolishing.

Methods 1 to 3 are suitable for studying defects in the as-grown
films only, while methods 4 to 6 suffer from the disadvantage that
defects are introduced in the process of thinning down a thick specimen. For most applications in metal physics it is necessary to take a
thin section from the interior of a bulk specimen after suitable treatment in such a way that the defect distribution in the thin foil is
likely to be characteristic of that of the bulk material, i.e. so that no
appreciable rearrangement of dislocations has occurred in the thinning process. Methods 7 and 8 are suitable methods, particularly for
metallic specimens, although ion bombardment is important for non-
metallic specimens.

An electropolishing technique was employed in early observations
of substructure in deformed aluminium by Heidenreich (1949),
while Hirsch et al. (1956) used the technique of beating followed by
etching in dilute hydrofluoric acid in their observations of dislocations
in this material. The latter technique, although simple to use, is of
limited value; electropolishing has proved to be the most useful
method of preparing thin specimens of many metals, although the
recipes vary from metal to metal. Fig. 3.1 illustrates an electropolishing technique, due to Bollmann (1956), for preparing thin foils
of austenitic stainless steels. The specimen foil is initially disc-shaped,
about 250 $\mu$m thick, supported on a stem (fig. 3.1($a$)). The foil is varnished with a non-conducting lacquer leaving a window about 2 cm
diameter on each side, and is made the anode in a 40% sulphuric
acid + 60% phosphoric acid bath at 60 °C. The cathode is a pair of
pointed stainless steel electrodes (fig. 3.1($b$)), varnished except at
the tips, and in the first stage polishing proceeds at about 9 V with the
cathode points close to the centre of the specimen until a small hole
appears as in fig. 3.1($c$). The points are then moved further away

from the disc and polishing is continued so as to produce a hole at
the edge of the disc (fig. 3.1(c)). Further polishing causes the edge
and central holes to coalesce, and pieces of the specimen cut with a

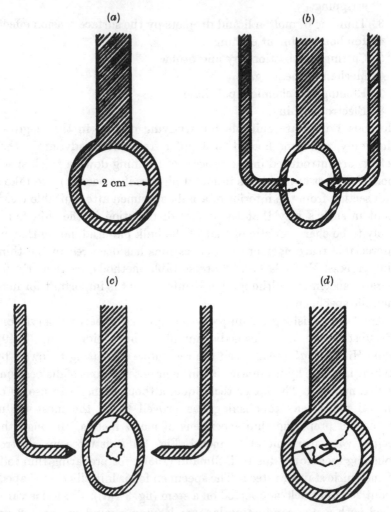

Fig. 3.1. Diagram illustrating Bollmann's electropolishing technique for stainless
steel. (Kelly and Nutting, 1964)

razor blade from the junction region (fig. 3.1(d)) are thin enough for
use in the electron microscope. To handle very thick specimens,
such as might be used in a tensile test, resort must be made to a
rough thinning process before electropolishing. Several methods

have been developed for this, including electrolytic jet planing (for plate specimens), the acid-string saw, and the spark cutting technique.

There are many variants of these techniques, and the interested reader should consult a reference book on the subject, such as Tegart (1956), Kelly and Nutting (1964), and Hirsch *et al.* (1965).

### 3.2.2 Electron microscopy

A modern high resolution electron microscope consists of an illuminating system, together with three electron lenses for magnifying the image of the specimen, known as the objective, intermediate, and projector lenses. In addition, the illuminating system usually contains two condenser lenses. The lenses are usually of the magnetic type, and this produces rotation of the image as well as magnification. The specimen is placed in the bore of the objective lens pole-piece and the image formed by the objective is magnified by the intermediate lens to form a second image, which in turn is magnified by the projector and viewed on a fluorescent screen. Total magnifications ranging from $\sim\times 200$ up to $\sim\times 100\,000$ can be obtained by varying the excitations of the lenses. The most useful magnification depends on the type of defect under observation. A typical magnification for observing dislocations is $\sim\times 20\,000$, since this gives a reasonable field of view on the viewing screen. The instrumental resolving power of a modern electron microscope is usually of the order of about 2–3 Å, but the resolution achieved in practice depends on many factors such as the state of perfection of the instrument and the nature of the specimen used. With metal foils a resolution in the range 10–20 Å is obtained, limited mainly by chromatic aberration and astigmatism due to the specimen itself. To obtain optimum penetration it is usual to work with as high an accelerating potential of the electrons as possible. With most commercial instruments this is 100 kV, although instruments have been constructed, and are commercially available with accelerating potentials up to 1 MV or greater (Dupouy and Perrier, 1962; Kobayashi *et al.*, 1964a, b); for recent developments see (1) Swann *et al.*, 1974, (2) Proceedings of the Symposium on High Voltage Electron Microscopy (EMCON 72) in *J. Microscopy* **97**, Parts 1 and 2 (1973), (3) *Proceedings of the Fifth European Congress on Electron Microscopy* (*EMCON 72*) published by the Institute of Physics).

The electron microscope can also be used to record electron diffraction patterns from small areas of the specimen during observation.

This is achieved by placing an aperture in the first image plane and by reducing the strength of the intermediate lens so that the back focal plane of the objective is focused on the viewing screen. The technique is known as selected area diffraction, and the diameter of the area selected may be as small as 0.5 $\mu$m. With crystalline specimens the diffraction pattern gives useful information about the crystallographic orientation of the specimen (after correcting for lens rotations mentioned above), and about Bragg reflections responsible for contrast effects (see section 3.3). It is therefore very important to record a diffraction pattern with micrographs to extract the maximum information from the image.

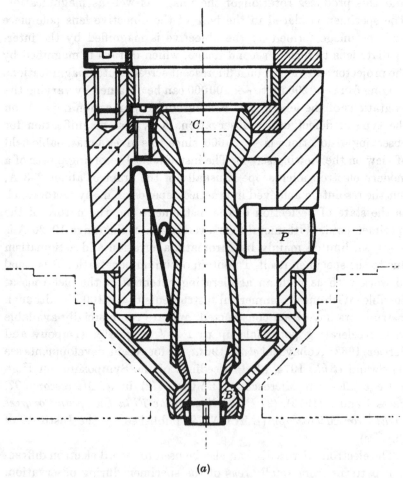

(a)

Mention should also be made at this point of the increasing interest in studying the energy loss spectrum of the electrons transmitted through the specimen in the electron microscope, since this contains much useful information on the electronic structure of the specimen. Instruments combining the normal facilities of electron microscopy and selected area electron diffraction with those of electron velocity analysers have been constructed and used to study a variety of

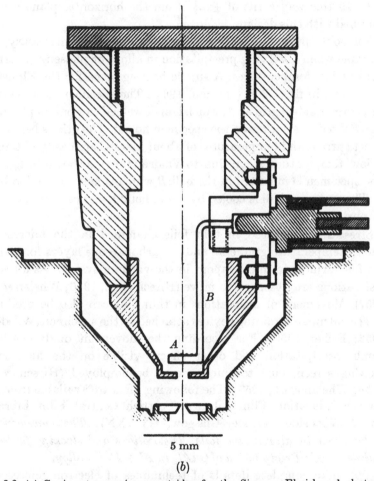

5 mm

*(b)*

Fig. 3.2. (*a*) Goniometer specimen cartridge for the Siemens Elmiskop 1 electron microscope. (Valdrè, 1962); (*b*) High temperature specimen cartridge for the Siemens Elmiskop 1 electron microscope. The specimen is mounted on a stainless steel grid at *A* supported by a pair of platinum legs *B*. The grid is heated by DC current. (Whelan, 1960)

problems (Watanabe and Uyeda, 1962; Castaing and Henry, 1962, 1964; Cundy et al., 1965; for applications see, for example, Cundy et al., 1968 and for instrumental review see Metherell, 1971).

It is also necessary to be able to tilt the specimen by a few degrees about two perpendicular axes in its plane to change the diffraction conditions. This is achieved with a goniometer-type specimen holder, an example of which (due to Valdrè, 1962) for the Siemens Elmiskop 1 electron microscope is shown in fig. 3.2(a). The specimen is mounted in a ball joint $B$ which is tilted by the drives $P$ and the lever pivoting at $C$. In this way a tilt of $\pm 22\frac{1}{2}°$ from the horizontal plane can be obtained with this design.

For some purposes heating or cooling stages are necessary, for example when observing precipitation in alloys or phase transformations at low temperatures. A simple heating stage for the Elmiskop 1 is shown in fig. 3.2(b) (Whelan, 1960). The specimen is supported on a stainless steel grid at $A$, which is spot welded to a pair of platinum legs $B$ fixed to a socket on the specimen holder. The grid is heated by direct current and temperatures of about 1000 °C are easily obtained. A low temperature stage due to Valdrè (1964) is shown in fig. 3.3. The specimen is mounted in the ball $B$ which is held in a nylon bush $N$. The specimen ball is cooled by liquid helium and can also be tilted by pushers $P$.

Stresses are induced in thin foils examined in the microscope due to temperature gradients and to carbonaceous layers formed on the foil surfaces from oil vapour in the vacuum system. As a result, dislocations are observed to move (Hirsch et al., 1956; Whelan et al., 1957). Movement of dislocations in thin foils can also be produced by special micro-straining devices attached to the specimen (Wilsdorf, 1958; Fisher, 1959). For recording the movement of dislocations, climb, precipitation and other effects visible on the fluorescent viewing screen, ciné techniques may be employed (Hirsch et al., 1956; Whelan et al., 1957). The following films are available from the Higher Education Film Library (Scottish Central Film Library, 16–17 Woodside Terrace, Glasgow, G3 7XN): *The movement of dislocations in aluminium foils*; *Dislocations and stacking faults in stainless steel*; *Precipitation of CuAl₂ in Al + 4%Cu alloy*.

For more complete details of techniques of electron microscopy and of special specimen stages reference should be made to standard treatises (Cosslett, 1950; Hall, 1953; Kay, 1965; Hirsch et al., 1965) and to recent conference reports listed on page 101.

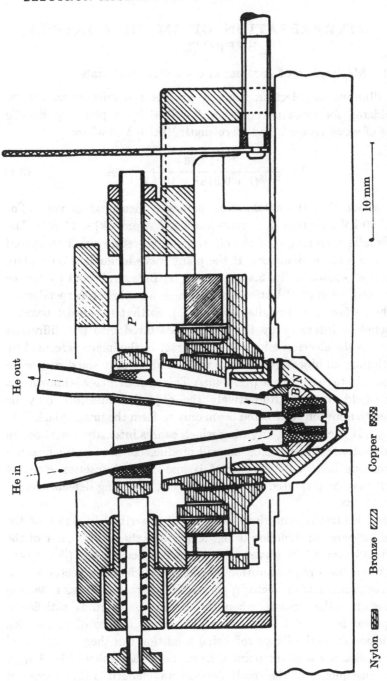

He out

He in

P

BN

10 mm

Nylon ⬚ Bronze ▨ Copper ▨

Fig. 3.3. Liquid helium temperature specimen cartridge for the Siemens Elmiskop 1 electron microscope. (Valdrè, 1964)

## 3.3 INTERPRETATION OF IMAGE CONTRAST EFFECTS

### 3.3.1 Mechanism of contrast in crystalline materials

The illuminating electron beam in an electron microscope can be considered for most purposes as described by a plane de Broglie wave of wave vector $\mathbf{k}$ and wavelength $\lambda(|\mathbf{k}| = \lambda^{-1})$ where

$$\lambda = \frac{12.26}{E^{1/2}(1 + 0.9788 \times 10^{-6} E)^{1/2}} \text{ Å} \tag{3.1}$$

and where $E$ is the accelerating potential measured in volts. For $E = 100$ kV equation (3.1) gives $\lambda = 0.037$ Å and $|\mathbf{k}| = 27$ Å$^{-1}$. The de Broglie wavelength of the electrons is therefore small compared with interatomic distances. If the plane wave is incident on a plate crystal as shown in fig. 3.4(a), it may be Bragg reflected by one or more sets of crystal lattice planes which are almost perpendicular to the surface (the so-called Laue case), so that the direct beam is depleted in intensity by the amount scattered into the diffracted beam. In the electron microscope, contrast in the image is formed by positioning an objective aperture so as to allow only the direct beam to contribute to the image (fig. 3.4(a)). This type of image is known as a bright-field image. Alternatively the objective aperture may be placed to allow one diffracted beam only to form the image (dark-field image). It is clear that anything which causes intensity variations in the direct and diffracted beams will produce contrast in the electron microscope image. Such contrast is known as diffraction contrast and it may be produced by various effects, including defects such as dislocations.

The diffraction conditions can also be described in terms of the Ewald sphere construction of fig. 3.4(b). $\mathbf{k}$ is the wave vector of the incident beam, $\mathbf{k}'$ the wave vector of the diffracted beam ($|\mathbf{k}'| = |\mathbf{k}|$), and when the Bragg condition is exactly satisfied the vector $\mathbf{k}' - \mathbf{k}$ is a reciprocal lattice vector $\mathbf{g}$ corresponding to the Bragg reflecting planes. In other words, when the Bragg condition is satisfied a reciprocal lattice point $\mathbf{g}$ lies on the Ewald sphere of radius $|\mathbf{k}|$. Deviation from the Bragg reflecting condition can then be described by the distance $\mathbf{s}$ of the point $\mathbf{g}$ from the Ewald sphere (fig. 3.4(c)). One consequence of the small electron wavelength is that angles of diffraction are very small. Since $|\mathbf{g}| \approx \frac{1}{2}$ Å$^{-1}$ for a low order reflection

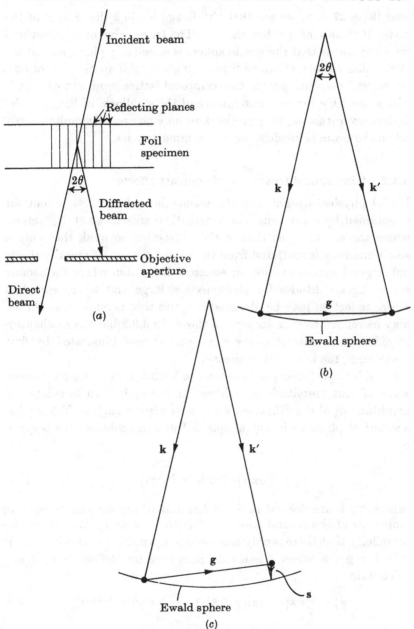

Fig. 3.4. (a) Diagram illustrating the mechanism of diffraction contrast. The thin foil specimen diffracts the electron beam and the diffracted beam is stopped by the objective aperture (bright-field image). The angle $\theta$ is of the order of 1° or 2°. (b) Ewald sphere construction: $\mathbf{g}$ is a reciprocal lattice vector. (c) Ewald sphere construction: Deviation from the exact Bragg reflecting condition is described by the distance $\mathbf{s}$ of the reciprocal lattice point $\mathbf{g}$ from the Ewald sphere

and $|\mathbf{k}| \approx 27$ Å$^{-1}$, we see that the Bragg angle $\theta$ (fig. 3.4) is of the order $10^{-2}$ radians, i.e. less than $1°$. The large value of $|\mathbf{k}|$ compared with $|\mathbf{g}|$ means that the Ewald sphere is effectively plane, so that the diffraction pattern obtained from a single crystal specimen is often a prominent plane of spots in the reciprocal lattice projected with little distortion. One further consequence of the smallness of Bragg angles is that several strong Bragg reflections may be excited simultaneously when the beam is incident along a symmetry axis.

### 3.3.2  Kinematical treatment of contrast effects

Useful physical insight into the mechanism of diffraction contrast is obtained by considering the kinematical treatment of diffraction, where the crystal is so thin or the diffraction so weak that only a small intensity is scattered from the incident beam. Naturally this is not a good approximation in electron diffraction where the atomic scattering amplitudes for electrons are large and where complete Bragg reflection may be obtained in quite thin crystals. The theory may be refined to take account of this as in 3.3.3 but the mechanism of diffraction contrast is the same and is best illustrated by first considering the kinematical theory.

Consider a perfect crystal as shown in fig. 3.5($a$) on which an electron wave of unit amplitude is incident. It is required to calculate the amplitude $\phi_{\mathbf{g}}$ of the diffracted wave at the lower surface. Taking due account of phase as in any simple diffraction problem, this is given by

$$\phi_{\mathbf{g}} \propto \sum_j \exp\left[-2\pi i(\mathbf{k}' - \mathbf{k})\cdot\mathbf{r}_j\right] \qquad (3.2)$$

where $\mathbf{k}'$, $\mathbf{k}$ are defined in fig. 3.4($c$) and where the sum is over all unit cells of the crystal situated at lattice points $\mathbf{r}_j$. We assume for simplicity that there is only one atom per primitive unit cell. Putting $\mathbf{k}' - \mathbf{k} = \mathbf{g} + \mathbf{s}$, where $\mathbf{s}$ is the deviation vector defined in fig 3.4($c$), we obtain

$$\phi_{\mathbf{g}} \propto \sum_j \exp\left[-2\pi i(\mathbf{g} + \mathbf{s})\cdot\mathbf{r}_j\right] = \sum_j \exp\left(-2\pi i\mathbf{s}\cdot\mathbf{r}_j\right). \qquad (3.3)$$

We have used the fact that $\mathbf{g}$ is a reciprocal lattice vector and $\mathbf{r}_j$ is a lattice vector so that $\mathbf{g}\cdot\mathbf{r}_j$ is an integer. The sum (3.3) may be replaced by an integral over the volume of the crystal. The $x$ and $y$ integration over the large surface area of the crystal leads to a $\delta$-

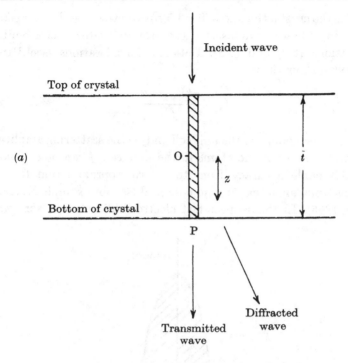

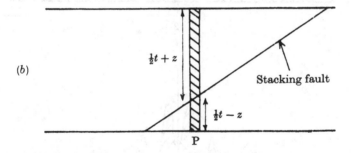

Fig. 3.5. (a) Section through a thin crystal illustrating the column of crystal for which the amplitude scattered is calculated. (b) Column of crystal for a stacking fault on an inclined plane. The waves scattered by the parts of the column on opposite sides of the fault differ in phase by an angle $\alpha = 2\pi\,\mathbf{g}\cdot\mathbf{R}$, where $\mathbf{R}$ is the displacement vector at the fault. (Whelan, 1959c)

function distribution in the co-ordinates $s_x$, $s_y$ and we eventually obtain

$$\phi_{\mathbf{g}} = \frac{\mathrm{i}\pi}{\xi_{\mathbf{g}}} \int_{-t/2}^{+t/2} \exp\left(-2\pi \mathrm{i} s z\right) \mathrm{d}z = \frac{\mathrm{i}\pi}{\xi_{\mathbf{g}}} \frac{\sin \pi t s}{\pi s} \qquad (3.4)$$

where $t$ is the crystal thickness (fig. 3.5(a)) and where we have replaced $s_z$ by $s$. In (3.4) we have inserted the proportionality factor omitted in equations (3.2) and (3.3). Detailed considerations (see Hirsch *et al.*, 1965) show that

$$\xi_g = \frac{\pi k V_c \cos \theta}{f} \tag{3.5}$$

where $V_c$ is the volume of the unit cell and $f$ is the scattering amplitude for electron waves of the atoms in the unit cell. $f$ can be evaluated with reasonable accuracy from the Born approximation for fast electrons using an actual atomic potential (see for example Mott and Massey, 1965). In this respect fast electron diffraction is somewhat

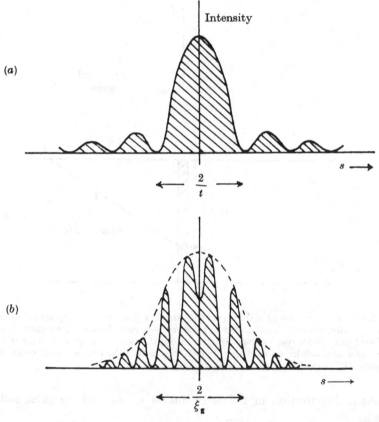

Fig. 3.6. Schematic diagram of the intensity distribution around a reciprocal lattice point, (a) on the kinematical diffraction theory, (b) on the dynamical diffraction theory. (Whelan, 1959c)

simpler than slow electron diffraction (as employed in the band theory of solids) where resort has to be made to 'pseudo-potentials' to calculate details of band structure. $\xi_g$ has the dimensions of length and is known as the extinction distance for the Bragg reflection $g$. Typical extinction distances are in the range 200–500 Å for most metals. The factor i occurring in (3.4) represents a phase change of $\pi/2$ which is well known in physical optics.

Equation (3.4) shows that the intensity distribution around the reciprocal lattice point corresponding to the Bragg reflection is proportional to

$$\frac{\sin^2 \pi ts}{(\xi_g s)^2} \tag{3.6}$$

and is in the form of a spike normal to the crystal surface as shown in fig. 3.6(a). The distribution has a central maximum of width $2/t$ with subsidiary maxima. Several contrast effects may now be interpreted qualitatively in terms of the ideas developed above. Figs. 3.7(a) and (b) show respectively effects known as thickness contours and bend extinction contours in an aluminium foil. With thickness contours the orientation of the crystal is fixed, i.e. $s$ in equation (3.6) (assumed non-zero) is constant, while $t$ varies, in this case in the vicinity of a pit on the crystal surface. As $t$ varies the width of the interference function in fig. 3.6(a) varies and the subsidiary maxima brush through the Ewald sphere causing intensity changes in the diffracted wave. Thus we see a number of concentric equal-thickness contours as in fig. 3.7(a). The depth periodicity of the fringes is given by

$$\Delta t = s^{-1}. \tag{3.7}$$

With bend contours the thickness $t$ is constant, and because the foil is buckled $s$ varies over the surface of the foil and the intensity distribution of fig. 3.6(a) is swept through the Ewald sphere causing a series of dark and light fringes. The kinematical theory therefore gives a reasonable qualitative account of fringes in perfect crystals, but is evidently quantitatively incorrect, since near the reflecting position $(s \to 0)$ equation (3.7) predicts that the fringe spacing becomes infinite, and in practice an upper limit to the spacing is observed.

Defects are made visible in the electron microscope through bending or discontinuities in the Bragg reflecting planes. This is

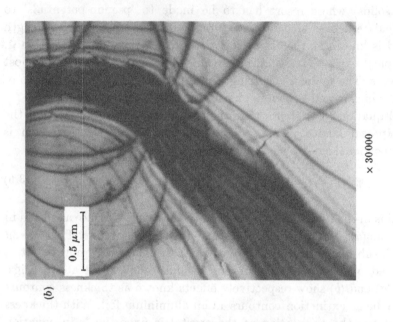

(b) × 30000

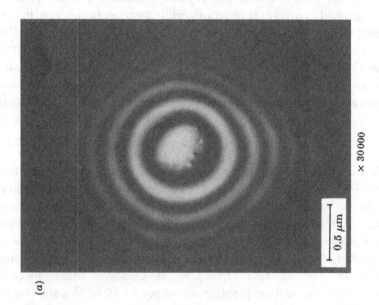

(a) × 30000

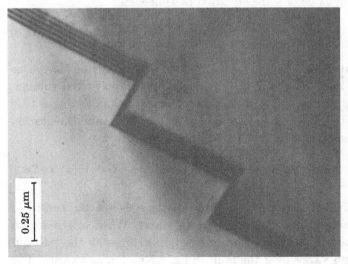

Fig. 3.7. Examples of interference fringes visible on electron micrographs of crystals. (a) Thickness extinction contours at a pit in an aluminium foil. (b) Bend contours in aluminium. (c) Thickness fringes at coherent twin boundaries in stainless steel. (d) Fringes at stacking faults in stainless steel. (Hirsch et al., 1960)

illustrated in fig. 3.5($b$) for a crystal containing a simple type of
planar defect known as a stacking fault. The stacking fault runs
through the foil on an inclined plane and the effect of the fault is to
displace the lower wedge crystal with respect to the upper by a
vector $\mathbf{R}$ which is not a lattice vector. Thus there is a phase difference
of $2\pi\mathbf{g}\cdot\mathbf{R}$ between waves diffracted on either side of the fault, and
it is not surprising that fringes are formed which are characteristic
of thickness fringes in either wedge crystal. An example of fringes at
stacking faults on {111} planes in a foil of stainless steel is shown in
fig. 3.7($d$), while fig. 3.7($c$) shows fringes of similar character at a
coherent twin boundary in the same material. Evidently therefore,
lattice displacements can cause a very marked contrast effect. To
treat the effect by the kinematical theory we consider a column
of crystal shown shaded in fig. 3.5($b$), and evaluate the amplitude $\phi_{\mathbf{g}}$
of the diffracted wave at the bottom of the column $P$. If the lattice
displacement is $\mathbf{R}$ at a point $z$ in the column, equation (3.3) becomes

$$\phi_{\mathbf{g}} \propto \sum_{j} \exp[-2\pi i(\mathbf{g}+\mathbf{s})\cdot(\mathbf{r}_{j}+\mathbf{R})] = \sum_{j} \exp(-2\pi i\mathbf{g}\cdot\mathbf{R} - 2\pi i\mathbf{s}\cdot\mathbf{r}_{j})$$
$$(3.8)$$

where we have neglected the small term in $\mathbf{s}\cdot\mathbf{R}$.

Proceeding as before, the equation (3.4) becomes

$$\phi_{\mathbf{g}} = \frac{i\pi}{\xi_{\mathbf{g}}} \int_{\text{Column}} \exp(-2\pi i\mathbf{g}\cdot\mathbf{R})\exp(-2\pi isz)\,dz. \qquad (3.9)$$

We see that $\phi_{\mathbf{g}}$ is essentially the Fourier transform of the phase
term $\exp(-2\pi i\mathbf{g}\cdot\mathbf{R})$ which is a function of position in the column.
Equation (3.9) may be evaluated for a stacking fault, where $\mathbf{R}$ is
zero above the fault and is a constant below the fault. Integration
of (3.9) (Whelan and Hirsch, 1957$a$) then gives

$$|\phi_{\mathbf{g}}|^2 = (\xi_{\mathbf{g}}s)^{-2}[\sin^2(\pi ts + \tfrac{1}{2}\alpha) + \sin^2\tfrac{1}{2}\alpha - 2\sin\tfrac{1}{2}\alpha\sin(\pi ts + \tfrac{1}{2}\alpha)\cos 2\pi zs]$$
$$(3.10)$$

where $\alpha = 2\pi\mathbf{g}\cdot\mathbf{R}$ and where $z$ is the distance below the centre of
the foil where the column intersects the fault. When $\alpha$ is not zero
or a multiple of $2\pi$, this equation predicts fringes as the position
of the column varies over the fault in fig. 3.5($b$), i.e. as $z$ varies. The
depth periodicity of the fringes is again given by equation (3.7).
When $\alpha$ is zero or a multiple of $2\pi$, equation (3.10) reduces to the
perfect crystal expression (3.6), and this illustrates the fact that when
the displacement of the fault does not produce any net offset in the

Bragg planes the fault is invisible. This is a useful criterion for investigating the displacement vector **R** at a fault if this is unknown. The fault is imaged in several Bragg reflections using a goniometer stage. Knowledge of the reflections in which the contrast vanishes is usually sufficient to give the displacement vector unambiguously. An example for faults in hexagonal AlN has been given by Drum (1965). For further applications see section 3.5.3.

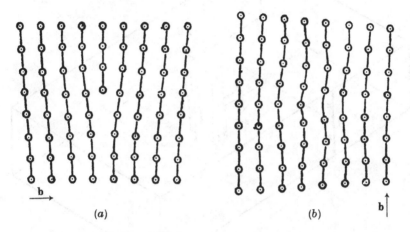

Fig. 3.8. Diagram showing the nature of atomic displacements near an edge dislocation: (a) The opposite sense of tilt of lattice planes normal to the Burgers vector on opposite sides of the dislocation; (b) Similar bending of planes containing the Burgers vector on opposite sides of the dislocation. (Howie, 1961)

Equation (3.9) may also be used to calculate the contrast from a crystal containing a dislocation by inserting the appropriate displacements **R**. The nature of the bending of the atomic planes near an edge dislocation is illustrated in fig. 3.8, while fig. 3.9 shows the situation for a screw dislocation line $AB$ parallel to the surface of a foil. The column of crystal $CD$ is deformed to the shape $EF$ by the presence of the dislocation. The displacement around the screw is given by

$$\mathbf{R} = \frac{b}{2\pi}\phi = \frac{b}{2\pi}\tan^{-1}\left(\frac{z}{x}\right) \qquad (3.11)$$

so that the quantity to be inserted in (3.9) is

$$2\pi \mathbf{g} \cdot \mathbf{R} = \mathbf{g} \cdot \mathbf{b}\tan^{-1}\left(\frac{z}{x}\right) = n\tan^{-1}\left(\frac{z}{x}\right). \qquad (3.12)$$

Since $\mathbf{g}$ is a reciprocal lattice vector and $\mathbf{b}$ is a lattice vector, the quantity $n = \mathbf{g} \cdot \mathbf{b}$ is an integer, which may take values $0, \pm 1, \pm 2$ etc. Values of the integral (3.9) can be obtained by computer or by graphical methods (see for example Hirsch *et al.*, 1960). Dislocation

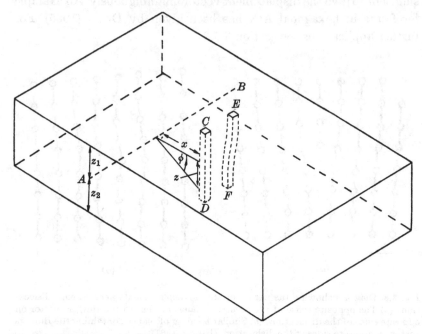

Fig. 3.9. Thin foil containing a screw dislocation $AB$. A column of crystal $CD$ in the perfect crystal is deformed to the shape $EF$ when the screw dislocation is present. (Hirsch *et al.*, 1960)

image profiles can then be obtained by calculating (3.9) as a function of position $x$ of the column in fig. 3.9. The results show:

1. The image of a screw dislocation lies to one side of the dislocation core depending on the deviation parameter $s$. This is a result of the fact that when the two phase factors in the integral (3.9) are of the same sign the integral oscillates rapidly about a small mean value, but when they are of opposite sign a large mean value is achieved. Since $\mathbf{R}$ changes sign on opposite sides of a screw dislocation, it follows that the contrast at a dislocation is asymmetrical about the core. Fig. 3.10 shows an example of this behaviour for dislocations in aluminium. The physical reason for the asymmetrical contrast in the case of an edge dislocation is evident from digrams like fig. 3.8(*a*), where it is seen that the rotation of lattice planes is of opposite sense

on opposite sides of the dislocation. Thus one side of the dislocation is tilted towards the reflecting position causing increased contrast, while the other side is tilted away from the reflecting position.

2. The peak of the image of a dislocation occurs where the phase of the integrand in (3.9) is almost stationary over an appreciable distance in the column. For large values of $s$ this occurs very close to

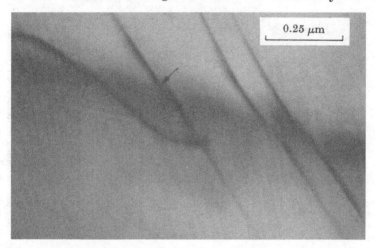

Fig. 3.10. Electron micrograph illustrating the one-sided nature of dislocation image contrast in aluminium. At the points marked with arrows, the contrast changes from one side of the dislocation to the other as the dislocation crosses the extinction contour. (Hirsch *et al.*, 1960)

the dislocation core, where the gradient of **R** is large. For a screw dislocation with $\mathbf{g} \cdot \mathbf{b} = 2$, it occurs at a distance from the core $x = (2\pi s)^{-1}$. The half-width of the image is $\Delta x = (\pi s)^{-1}$. These results show that, by using large $s$ values, the image can be made very narrow with the peak close to the core. The intensity in the diffracted beam varies as $s^{-2}$, so that with large $s$ the contrast in the bright-field image is low. However a dark-field image, although weak, gives good contrast if exposed sufficiently. This is the theoretical basis of the weak-beam technique for high resolution imaging of dislocations. The development of this technique in recent years has been greatly facilitated by the advent of electromagnetic beam-tilting facilities in electron microscopes. To obtain a dark-field image of good quality it is necessary for the diffracted beam to pass axially down the instrument, and this requires the incident beam to be inclined to the axis. The method was first demonstrated by Cockayne *et al.* (1969). Some applications of the technique are mentioned in section 3.5.2. For a

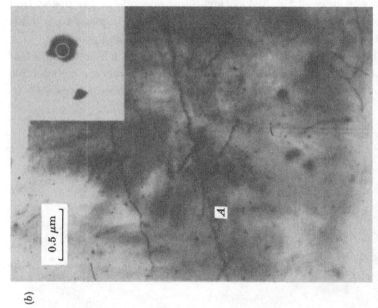

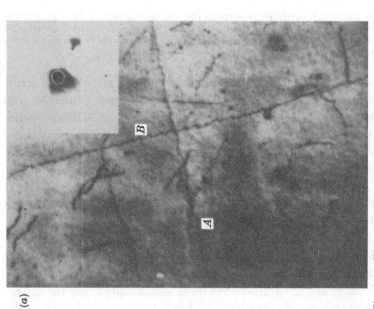

Fig. 3.11. Electron micrographs and diffraction patterns illustrating the vanishing of dislocation contrast: (a) Both dislocations A and B are visible in 020 Bragg reflection; (b) Dislocation A is visible but B is invisible for 2$\bar{2}$0 reflection. (Hirsch et al., 1960)

review see the Proceedings of the Symposium on Applications of the Weak-Beam Technique in Materials Science (EMCON 72) in *J. Microscopy* **98**, Part 1 (1973).

3. Edge dislocations may give images similar to those of screws, except that the image may be rather wider.

4. For certain reflections $\mathbf{g} \cdot \mathbf{b}$ may vanish. In this case no diffraction contrast is produced, i.e. the dislocation is invisible. Since $\mathbf{g}$ is normal to the reflecting planes, the above condition implies that the Burgers vector of the dislocation lies in the reflecting planes, i.e. that displacements parallel to the reflecting planes cause no contrast. This result is exact for screw dislocations and also holds approximately for edge dislocations in many cases. For an edge dislocation viewed parallel to the slip plane and parallel to the Burgers vector (i.e. $\mathbf{g} \cdot \mathbf{b} = 0$), contrast can be caused by displacements normal to the slip plane, illustrated in fig. 3.8($b$). This contrast is symmetrical about the centre of the dislocation, since the bending of lattice planes on either side is the same. Examples have been given for dislocation loops in zinc by Howie and Whelan (1962), and for defects in stainless steel by Silcock and Tunstall (1964). Fig. 3.11 is an example of vanishing dislocations in a quenched Al–4wt.%Cu alloy. The long dislocations visible are helices formed by vacancy climb from screw dislocations running in [$\bar{1}$10] and [110] directions. The plane of the foil is close to (001). Now the diffraction pattern (inset) from the region of fig. 3.11($a$) shows that the 020 reflection is producing the contrast. Since the helical dislocations $A$ and $B$ have Burgers vectors along their axes, the Burgers vectors of the long dislocations are known to be along their lengths. Therefore $|\mathbf{g} \cdot \mathbf{b}| = 1$ for the dislocations $A$ and $B$ and both are visible in fig. 3.11($a$). Fig. 3.11($b$) shows the same region after tilting the specimen. The 020 reflection has been removed and the $2\bar{2}0$ reflection has appeared. This reflection has $\mathbf{g} \cdot \mathbf{b} = 0$ for dislocation $B$ and $|\mathbf{g} \cdot \mathbf{b}| = 2$ for $A$. Dislocation $B$ therefore vanishes in agreement with theory. The $\mathbf{g} \cdot \mathbf{b} = 0$ criterion provides a most useful method of determining the Burgers vector of a dislocation (see section 3.5.3).

For a more detailed account of the theory outlined in this section the reader should consult Hirsch *et al.* (1965).

### 3.3.3 Dynamical theory of contrast effects

The kinematical theory is in many respects unsatisfactory since it neglects to take account of the equilibrium which must exist between

direct and diffracted waves in a crystal. To see this we note that if
$s = 0$ in (3.6) the diffracted beam can exceed unit intensity if $t > \xi_g/\pi$,
and that the fringe spacing given by equation (3.7) tends to infinity
as $s$ tends to zero. Obviously this is an absurd state of affairs, since the
diffracted beam is now stronger than the incident beam. To improve

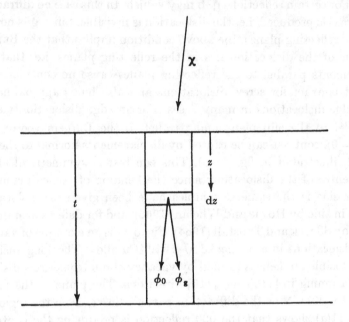

Fig. 3.12. Direct ($\phi_0$) and diffracted ($\phi_g$) waves propagating through an element d$z$ of
a column of crystal. $\chi$ is the wave vector of the incident electron beam.

the theory we consider the equations describing the equilibrium
between the direct ($\phi_0$) and diffracted ($\phi_g$) wave amplitudes in the
column of crystal in fig. 3.12. It can be shown (Howie and Whelan,
1961) that the equations are

$$
\left.
\begin{aligned}
\frac{d\phi_0}{dz} &= \frac{i\pi}{\xi_o}\phi_0 + \frac{i\pi}{\xi_g}\phi_g \\
\frac{d\phi_g}{dz} &= \frac{i\pi}{\xi_g}\phi_0 + i\left(\frac{\pi}{\xi_o} + 2\pi s\right)\phi_g.
\end{aligned}
\right\}
\tag{3.13}
$$

The first equation effectively states that the change in direct wave
amplitude in an element d$z$ of the column of fig. 3.12 is partly due
to forward scattering of the direct wave and partly due to Bragg

scattering of the diffracted wave. The second equation has a similar interpretation. Now $\xi_0$ (which also has the dimensions of length) can be removed by making a phase transformation

$$\phi_0' = \phi_0 \exp\left(-i\pi z/\xi_0\right), \quad \phi_g' = \phi_g \exp\left(-i\pi z/\xi_0\right).$$

This simply represents the change in wavelength (or refractive index effect) due to forward scattering, an effect which is well known in physical optics. We then obtain

$$\left. \begin{aligned} \frac{d\phi_0'}{dz} &= \frac{i\pi}{\xi_g} \phi_g' \\ \frac{d\phi_g'}{dz} &= \frac{i\pi}{\xi_g} \phi_0' + 2\pi is\, \phi_g'. \end{aligned} \right\} \tag{3.14}$$

These equations may now be solved with the matching conditions $\phi_0'(0) = 1$, $\phi_g'(0) = 0$ at the top surface in fig. 3.12, and we obtain (dropping further use of primes)

$$\phi_0(t) = \cos\left(\frac{\pi t}{\xi_g} \sqrt{(1 + w^2)}\right) - \frac{iw}{\sqrt{(1 + w^2)}} \sin\left(\frac{\pi t}{\xi_g} \sqrt{(1 + w^2)}\right) \tag{3.15a}$$

$$\phi_g(t) = \frac{i \sin\left(\dfrac{\pi t}{\xi_g} \sqrt{(1 + w^2)}\right)}{\sqrt{(1 + w^2)}} \tag{3.15b}$$

where $w = \xi_g s$ is a dimensionless parameter which denotes deviation from the reflecting condition. We note from equations (3.15) that $|\phi_0|^2 + |\phi_g|^2 = 1$, i.e. intensity is conserved, and that (3.15$b$) is the same as (3.4) if we replace $s$ in that equation by

$$s_{\text{eff}} = (s^2 + \xi_g^{-2})^{1/2}. \tag{3.16}$$

Thus we see that according to the dynamical theory $s_{\text{eff}}$ cannot be zero but has a minimum value equal to $\xi_g^{-1}$. We note also that (3.15$b$) removes the previous difficulty of $|\phi_g|^2$ being greater than unity when $t > \xi_g/\pi$ at $s = 0$. The form of the intensity distribution around a reciprocal lattice point is given by

$$\frac{\sin^2 \pi t s_{\text{eff}}}{(\xi_g s_{\text{eff}})^2} \tag{3.17}$$

and is of the general form illustrated in fig. 3.6($b$); it is to be compared with equation (3.6). The curve does not necessarily have a maximum at $s = 0$. While the shape of the curve differs in detail from that of fig. 3.6($a$), the qualitative conclusions drawn about fringes at thickness and bend contours from kinematical theory remain unchanged. Equations (3.15) also predict that the intensity of the direct and diffracted waves oscillate with depth $t$ in a crystal. At the Bragg reflecting position the form of the oscillation is as shown in fig. 3.13($b$);

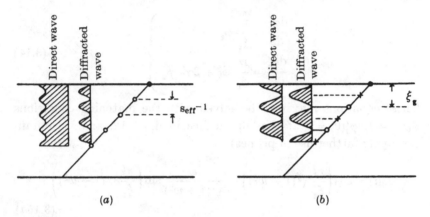

(a)                                        (b)

Fig. 3.13. Schematic diagrams of the depth intensity oscillations of the direct and diffracted waves: ($a$) Kinematical theory; ($b$) Dynamical theory. (Whelan and Hirsch, 1957$a$)

the intensities of direct and diffracted waves oscillate in antiphase between zero and unity with a periodicity $\xi_g$ in depth. Away from the Bragg position the oscillations are as shown in fig. 3.13($a$), and have a depth periodicity $s_{\text{eff}}^{-1}$. This tends to $s^{-1}$ for large $s$, in agreement with the kinematical theory. The simple physical interpretation of the extinction distance $\xi_g$ is now easily seen from fig. 3.13($b$).

The formulation of the dynamical theory as outlined here follows an approach similar to that used by Darwin (1914) for X-ray diffraction, and while simple and instructive, certain important features of the problem are not easily demonstrated by this approach. An alternative (and equivalent) formulation of the theory uses the approach via the Schrödinger equation of the fast electron in the periodic potential of the crystal. The solution is then obtained in terms of Bloch waves, and certain important effects arise owing to the symmetry of the Bloch waves. It turns out that there are two possible Bloch waves which can propagate in the crystal with the forms shown in fig. 3.14 at the Bragg reflecting position. One of these waves is localised on the reflecting planes of atoms, while the other is localised between the planes. Now inelastic scattering can occur at

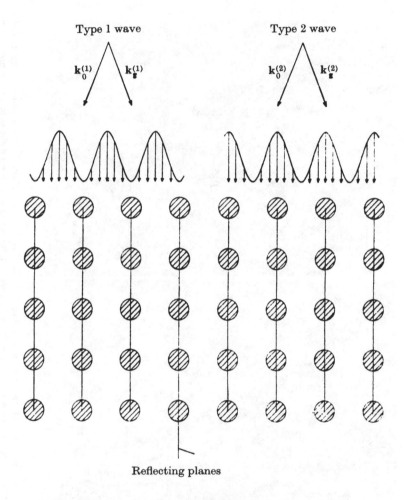

Fig. 3.14. Schematic diagram showing the two types of Bloch wave which propagate parallel to the reflecting planes at the exact Bragg position. (Hashimoto *et al.*, 1962)

the atoms by mechanisms such as interband electronic excitation and phonon scattering. This type of scattering can remove the electron completely from the image by scattering outside the objective aperture of the microscope and can hence contribute to absorption. It is clear that if the inelastic scattering is localised in the shaded

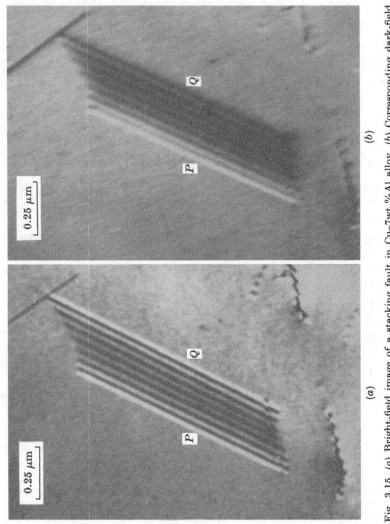

(a)                                                    (b)

Fig. 3.15. (a) Bright-field image of a stacking fault in Cu–7wt.%Al alloy. (b) Corresponding dark-field image in a strong 111 reflection. Note the symmetry of the fringes in the bright-field image and their asymmetry in the dark-field image. (Hashimoto et al., 1962)

regions of the atoms in fig. 3.14, the type 2 wave will be absorbed more than the type 1 wave. Thus we have the possibility of 'channelling' or easy penetration of one Bloch wave. This effect is called anomalous absorption and it is the electron counterpart of the well known Borrmann effect in X-ray diffraction. This effect enables electron microscopy to be performed on thicker specimens of crystalline material than would otherwise be possible. The preferential absorption of one Bloch wave also leads to the result that the visibility of thickness fringes decreases in thick regions of crystals, as shown in fig. 3.7(a). A discussion of anomalous absorption effects has been given by Hashimoto et al. (1962), while the mechanisms of absorption have been studied by Yoshioka (1957), Hall and Hirsch (1965), and Whelan (1965a, b).

The dynamical theory of image contrast can also be extended to cover defects such as stacking faults, dislocations and small inclusions in crystals. The effect of lattice displacements due to a defect can be incorporated in the theory in the same way as in the kinematical theory, i.e. by inserting the appropriate displacements $\mathbf{R}$ in a column of crystal near the defect. It can be shown (Howie and Whelan, 1961) that equations (3.14) are now modified by the substitution

$$s \rightarrow s + \frac{\mathrm{d}}{\mathrm{d}z}(\mathbf{g} \cdot \mathbf{R}) \tag{3.18}$$

and the resulting equations can be integrated by computer in particular cases. For stacking faults, the solution can be evaluated analytically (Whelan and Hirsch, 1957a, b), and anomalous absorption effects can be included (Hashimoto et al., 1962). One interesting effect at images of stacking faults when absorption occurs is that the dark-field image is no longer symmetrical with respect to the top and bottom surfaces of the foil ($P$ and $Q$ in fig. 3.15), as is the case for the bright-field image. The visibility of the fringes is also lower at the centre of the fault. An example is shown in fig. 3.15 for Cu–7wt.%Al alloy. It is possible to use this effect to determine the type of fault present, i.e. to determine whether the fault is intrinsic or extrinsic (Hashimoto et al. 1962; see also section 3.5.3). For the case of dislocations reference should be made to papers by Howie and Whelan (1961, 1962), Silcock and Tunstall (1964), for the case of inclusions, to papers by Ashby and Brown (1963a, b), and for a general discussion to the book by Hirsch et al. (1965).

## 3.4  OBSERVATIONS OF DEFECTS IN VARIOUS METALS

We now briefly review some observations made of defects in several metals after various treatments.

### 3.4.1  Polycrystalline metals

*Aluminium.* The first metals examined by transmission electron microscopy were polycrystalline. Heidenreich (1949) studied aluminium, lightly deformed and thinned by electropolishing, and first observed the substructure which was also detected by microbeam X-ray diffraction (Gay et al., 1954). Fig. 3.16 shows the substructure in beaten and etched aluminium foil (Whelan, 1959a). The subgrains or cells are 1 to 2 $\mu$m in diameter with misorientations across the subgrain walls of $\sim$1 or 2°. Most of the dislocations are arranged in the subgrain walls which are small angle boundaries of the types discussed by Burgers (1939), Frank (1955) and others. From the observed misorientations the dislocation density in the walls is $\sim$10$^{10}$cm$^{-2}$. Fig. 3.16 shows a crossed-grid of dislocations at $A$, which probably corresponds to a twist boundary on (100) (Burgers, 1940). Fig. 3.16 also shows at $B$ a cross-slip trace left by a screw dislocation which has moved on $\{111\}$ planes in the thin foil during observation. When a dislocation moves in the foil a step at the surface is often produced as shown in fig. 3.17; however, owing to the presence of an oxide film on the surfaces of the foil the formation of the step is hindered and the resulting strain gives rise to diffraction contrast effects so that the trace of the slip plane is made visible (Howie and Whelan, 1962). The slip-trace at $B$ changes direction through a right angle in the subgrain, showing that the dislocation cross-slips to another plane. This represents direct proof of the Mott–Frank screw dislocation mechanism of cross-slip (Mott, 1951). Cross-slip of screw dislocations is observed frequently in aluminium, as expected for a metal of high stacking fault energy (Seeger, 1955a; Schoeck and Seeger, 1955). The ease with which cross-slip occurs in aluminium is probably partly responsible for the ability of the dislocations to arrange themselves in the low energy configurations of the cell walls.

*Stainless steel.* Dislocation arrangements in polycrystalline specimens of austenitic (f.c.c.) stainless steel (18wt.% Cr, 8wt.% Ni were first reported by Bollmann (1956) and later by Whelan et al. (1957). The

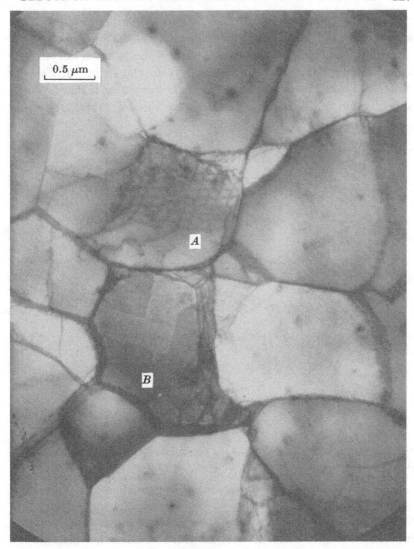

Fig. 3.16. Subgrain structure in beaten aluminium foil, thinned by etching in dilute HF solution. (Whelan, 1959a)

grain size was ~10 $\mu$m, and at deformations of the order of 1 or 2 per cent by rolling, the dislocations are observed to form tangled three-dimensional networks in the interior of the grains (fig. 3.18). There is no tendency to form a substructure even at high deformation. The dislocation density in the tangles varies from ~$10^9$ to ~$10^{11}$ cm$^{-2}$ for deformations ranging from 2 to 30 per cent.

A prominent feature of the observations at low deformations is the appearance of piled-up groups of dislocations at the grain boundaries. For example, in fig. 3.19 the spacing of the dislocations in the piled-up groups conforms at least approximately to the theory of Eshelby *et al.* (1951) (see also chapter 2). A further observation is that no cross-slip of moving dislocations occurs. The absence of cross-slip and

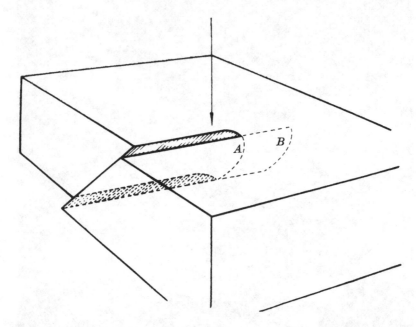

Fig. 3.17. Schematic diagram showing the formation of slip steps on the surface of a thin foil by a moving dislocation. If the surface oxide film is very strong the formation of the steps may be inhibited leaving strained regions (*A*, *B*) at the surface. The arrow indicates the direction of view in the electron microscope. (Hirsch, 1959*a*)

the formation of piled-up groups is as expected for a metal of low stacking fault energy, where the ribbon-width of extended dislocations and the activation energy of cross-slip are large. This behaviour, and also the absence of substructure formation, is in contrast to the behaviour of aluminium.

The dislocation networks observed in stainless steel are formed by the interaction of piled-up groups with secondary dislocations. Figs. 3.20 and 3.21 illustrate some of the interactions observed in the networks. In fig. 3.20 a group of dislocations piled-up at the grain boundary has interacted with secondary dislocations at *A*, *B*, *C* and elsewhere. An analysis of the interaction in terms of Thompson's

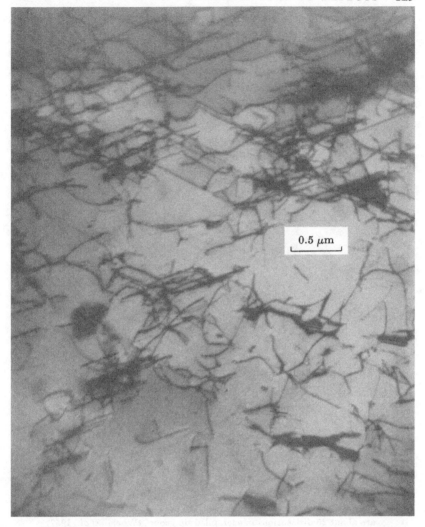

Fig. 3.18. Complex networks of dislocations in stainless steel deformed 2 per cent by rolling

notation (1953) is given in fig. 3.22. The primary piled-up dislocation (shown here as a ribbon $D\alpha - \alpha C$) on plane $(a)$ interacts over a short length with a secondary screw dislocation $(C\delta - \delta B)$ on plane $(d)$. The interaction in terms of partial dislocation Burgers vectors is $\alpha C + C\delta + \delta B = \alpha B$, as shown in fig. 3.22$(a)$. The network then rearranges under the action of line tension to give the arrangement of 3-fold extended and contracted nodes shown in fig. 3.22$(c)$. The extended nature of alternate nodes in fig. 3.20 is clearly visible; such nodes are very useful for estimating the stacking fault energy (see section 3.5.2). The stacking fault energy estimated in this way is $\sim13$ erg cm$^{-2}$. This is low in comparison with that of aluminium which is thought to be about 200 erg cm$^{-2}$ (see also chapter 2).

Fig. 3.19. Piled-up groups of dislocations at a grain boundary in stainless steel. There are about 25 dislocations in each piled-up group. (Whelan *et al.*, 1957)

Figure 3.21 shows another type of interaction frequently observed The interaction occurs along short lengths (LC) to form Lomer–Cottrell locks, which in this case cannot rearrange to form a hexagonal net because of the sessile nature of the lock. These and other interactions in the networks have been discussed by Whelan (1959b).

*Other metals.* Examples of dislocation arrangements in other polycrystalline metals are shown in figs. 3.23 to 3.25. Fig. 3.23 shows dislocations in α-brass after 16 per cent deformation. This material

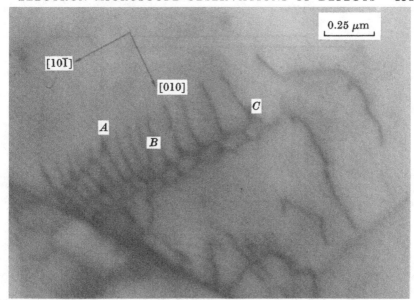

Fig. 3.20. Hexagonal network of dislocations in stainless steel, formed by interaction of dislocations with Burgers vectors at 120° (see fig. 3.22). Note the extended and contracted nature of the three-fold nodes. (Whelan, 1959b)

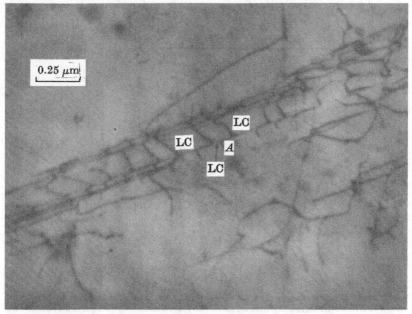

Fig. 3.21. Interactions of dislocations in stainless steel. A piled-up group of dislocations has interacted with several long dislocations to produce short lengths of Lomer–Cottrell barrier (LC). (Whelan, 1959b)

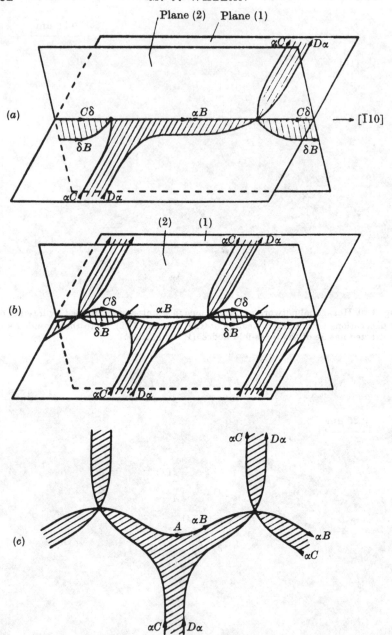

Fig. 3.22. Schematic diagram of the interaction of ribbon dislocations with Burgers vectors at 120°. (a) shows the first stage of the interaction: $C\delta$ and $\delta B$ are absorbed by $\alpha C$. In (b), $\alpha B$ pulls down and the nodes indicated with arrows are driven along by the line tension of $D\alpha\alpha C$. (c) shows the final configuration. Note the extended and contracted nodes (compare with fig. 3.20). (After Whelan, 1959b)

has a low stacking fault energy, and the dislocation arrangement is similar to that in stainless steel, i.e. tangled networks are formed and there is no tendency to form substructure. With metals of increasing stacking fault energy tangles of dislocations are observed bounding

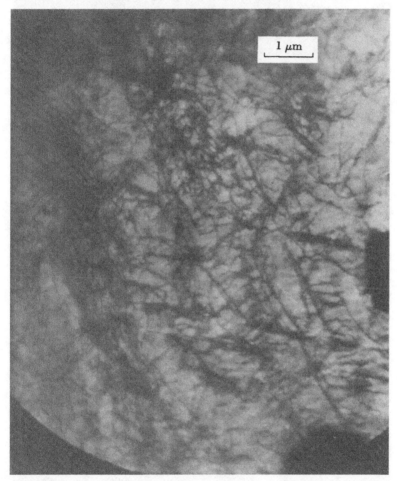

Fig. 3.23. Irregular arrangement of dislocations in α-brass deformed 16 per cent. (Whelan, 1959a)

regions of lower dislocation density (figs. 3.24 and 3.25). This corresponds to a poorly developed substructure in which the arrangement of dislocations in the cell walls is not as perfect as in the case of aluminium. Dislocation arrangements in other metals, including hexagonal and body-centred metals, have been reported by Tomlinson (1958) and others. A review is given Whelan (1959a).

In recent years considerable interest has centred round the study of dislocations in single crystals of metals and their correlation with stress–strain behaviour. The electron microscope has played an

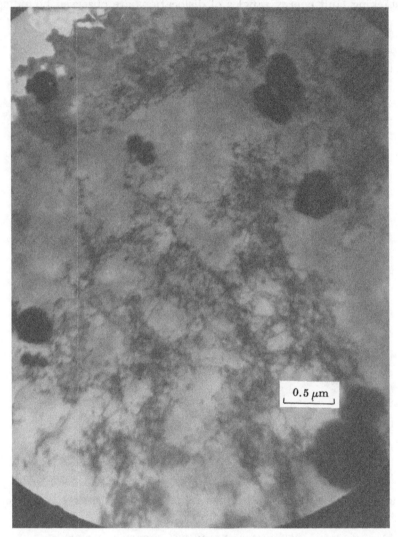

Fig. 3.24. Arrangement of dislocations in copper deformed 10 per cent in tension. The first stage of cell structure formation is visible. (Whelan, 1959a)

important part in the elucidation of dislocation behaviour in single crystals, but since the subject is reviewed in chapter 5 no further account will be given here.

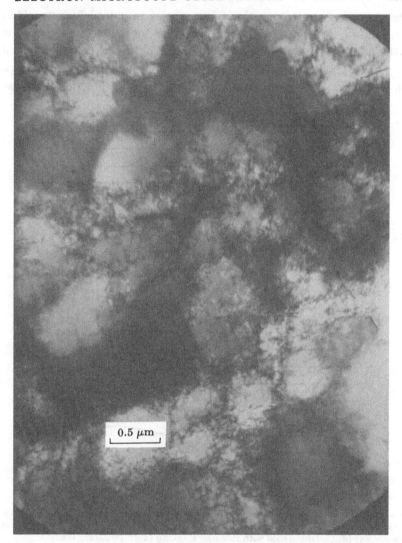

0.5 μm

Fig. 3.25. Cell structure formation in nickel deformed 25 per cent. (Whelan, 1959a)

## 3.4.2 Quenched and irradiated metals

*Quenched metals and alloys.* When a metal is quenched from high temperatures or subjected to bombardment by high energy nuclear particles, vacancies and interstitials are produced and may be trapped in the lattice. Quenching produces mainly vacancies because of their lower energy of formation, while irradiation may produce both types of point defect (for review see Cottrell, 1956). For example, the degree of supersaturation of vacancies produced by quenching aluminium from near the melting point to room temperature is ~$10^9$. Frank (1950), Seitz (1950), and Kuhlmann-Wilsdorf (1958) suggested that the excess vacancies would coagulate to form loops of dislocation as shown schematically in fig. 3.26. Direct collapse

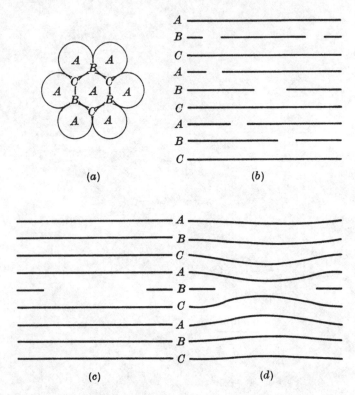

Fig. 3.26. Schematic diagrams illustrating dislocation loop formation by vacancies in a face-centred cubic metal: (a) Stacking sequence; (b) Random vacancies; (c) Disc of vacancies; (d) Collapsed disc with intrinsic stacking fault. (Whelan, 1963)

of a disc of vacancies will produce a loop of Frank partial dislocation containing an intrinsic stacking fault (fig. 3.26(d)). However, if the stacking fault energy is high the faulted loop may be converted by shear to a prismatic loop of whole dislocation line (see also section 2.11).

Figures 3.27 and 3.28 show examples of both types of loop in quenched aluminium. Fig. 3.27 (Hirsch et al., 1958) shows unfaulted loops in 99.995% pure aluminium after quenching from ~600 °C in iced brine and ageing at room temperature. It is seen that there is a denuded zone ~1 $\mu$m wide on either side of the grain boundary, due to the fact that the grain boundary acts as a sink for vacancies preventing nucleation of loops. The fractional vacancy concentration required

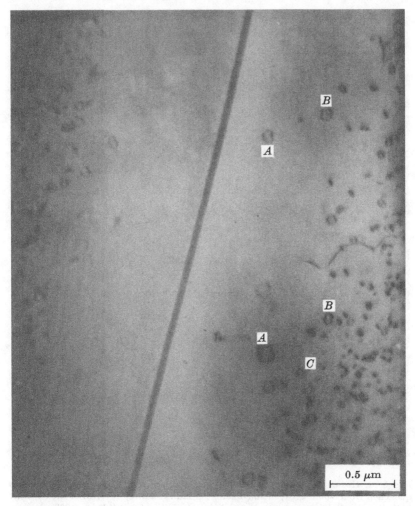

Fig. 3.27. Dislocation loops without stacking faults in aluminium quenched from ~600°C into iced brine. Note the denuded zone near the grain boundary. (Hirsch *et al.*, 1958)

to form the loops is estimated from the size and density of the loops to be ~$10^{-4}$ at 600°C, in good agreement with that calculated from the activation energy of formation of vacancies in aluminium (1.3 eV). Fig. 3.28 shows large faulted loops, with typical stacking fault fringe contrast, in zone-refined aluminium quenched from 600 °C (Cotterill and Segall, 1963). The loops are much larger than those in the 99.995% pure material, but their density is smaller, so that the vacancy concentration is again $\sim 10^{-4}$. Cotterill and Segall (1963)

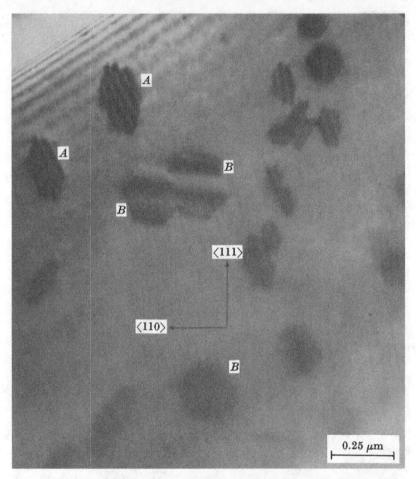

Fig. 3.28. Hexagonal sessile loops with stacking faults in zone refined aluminium quenched from ~600 °C. A, inclined stacking faults showing reversal of the contrast in the fringes. B, loops more nearly parallel to the foil surface. (Whelan, 1963)

showed that the type of loop observed after quenching aluminium was sensitive to quenching history, quenching temperature and the presence of impurities.

In gold quenched from 1000 °C, vacancies are observed to coagulate to form stacking fault tetrahedra (Silcox and Hirsch, 1959a) as shown

in fig. 3.29. The formation of stacking fault tetrahedra can be understood as follows. Imagine a triangular sessile loop of Burgers vector
$\alpha A$ occupying the area $CDB$ of the {111} face of Thompson's tetra-

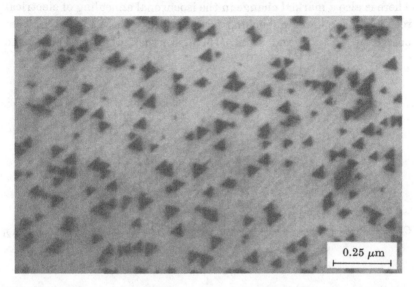

Fig. 3.29. Stacking fault tetrahedra in gold quenched from 1000 °C. (Cotterill, 1961)

hedron in fig. 3.30($b$). The Burgers vector $\alpha A$ of the Frank sessile
dislocations along $CD$, $DB$ and $BC$ may dissociate as follows on
planes $\beta$, $\gamma$ and $\delta$:

$$\alpha A \rightarrow \alpha\beta + \beta A$$
$$\alpha A \rightarrow \alpha\gamma + \gamma A$$
$$\alpha A \rightarrow \alpha\delta + \delta A.$$

$\alpha\beta$, $\alpha\gamma$ and $\alpha\delta$ are stair-rod dislocations at acute bends in the stacking
fault, while $\beta A$, $\gamma A$ and $\delta A$ are Shockley partials. The Shockley
partials bow out and attract each other on planes $\beta$, $\gamma$ and $\delta$ to form
stair-rod dislocations along the edges $AC$, $AD$ and $AB$, according to
reactions like

$$\beta A + A\gamma \rightarrow \beta\gamma \text{ etc.}$$

Thus a tetrahedral stacking fault defect is formed. It is likely, however, that the defect may grow in tetrahedral form from a very small
size by a mechanism involving jog lines on the faces of the tetrahedron (Silcox and Hirsch, 1959$a$).

Cotterill (1961) and Cotterill and Segall (1963) have shown that gold quenched from lower temperatures (~800 °C) contains no tetrahedra. Instead, a high density of small black spot defects appear. There is also a marked change in the isochronal annealing of electrical resistivity for lower quenching temperatures. These results are discussed by Cotterill and Segall (1963). Further investigations of the

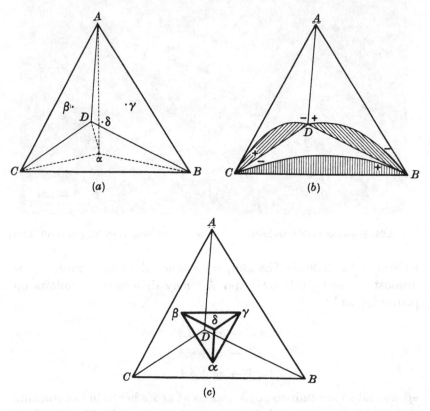

Fig. 3.30. Diagram illustrating the formation of a stacking fault tetrahedron. See text for detailed discussion. (Silcox and Hirsch, 1959a)

behaviour of quenched copper, silver, gold and alloys have been made by Clarebrough et al. (1964), Loretto et al. (1965, 1966), Clarebrough et al. (1966) and Segall et al. (1966).

Quenched Al–4wt.%Cu alloys have been examined by Thomas and Whelan (1959) and by Smallman et al. (1959), and the latter authors have also examined Al–20wt.%Ag alloys. Loops and helical dislocations are observed in both alloys. Fig. 3.31 is an example of helical dislocations in Al–4wt.%Cu alloy quenched from 540°C. Quenched-in vacancies migrate to screw dislocations and cause them to climb into helices (see also section 2.29). The formation of helical dislocations is a common observation in alloys of aluminium with copper, silver and magnesium, and their presence can account for precipitation patterns observed on ageing alloys (Thomas and Nutting, 1955).

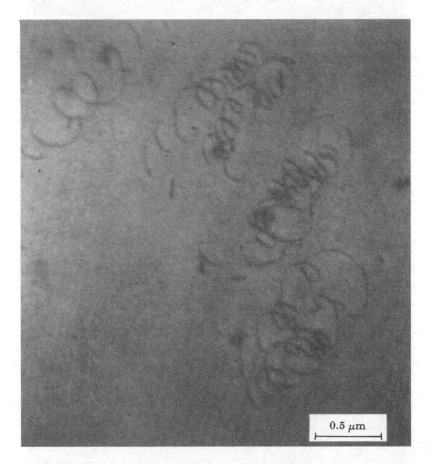

0.5 μm

Fig. 3.31. Helical dislocations in Al–4wt.%Cu alloy quenched from 540 °C. The helices are formed from screw dislocations by vacancy climb. (Thomas and Whelan, 1959)

*Irradiated metals.* The electron microscope has been used extensively to study the irradiated state of various metals. The specimen is usually irradiated in the bulk state in a reactor or accelerator and is subsequently thinned for examination. Silcox and Hirsch (1959*b*) first reported results on copper irradiated to various dosages with fast neutrons (>1 MeV), and fig. 3.32 shows one of their specimens irradiated to a large dose (~$1.4 \times 10^{20}$ neutrons cm$^{-2}$). Debris in the form of dislocation loops and black spot defects is visible. Silcox and Hirsch tentatively identified the loops as vacancy loops formed

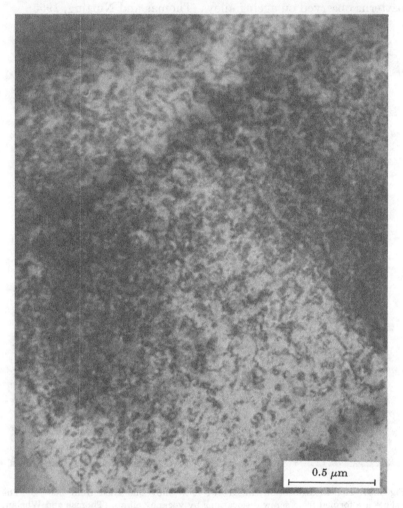

0.5 μm

Fig. 3.32. Dislocation loops and debris in copper irradiated with neutrons to a dose ~$1.4 \times 10^{20}$ neutrons cm$^{-2}$. (Silcox and Hirsch, 1959b)

at displacement spikes when the interstitials are shot away as dynamic crowdions, and they attempted to account for the irradiation-hardening on the basis of this loop model. This has been criticised on the grounds that the hardening expected from the loops is too small

(Makin *et al.*, 1961). Barnes and Mazey (1960) suggested that the loops observed in α-irradiated copper were in fact interstitial loops, and this was subsequently confirmed by a diffraction contrast analysis of loops in α-irradiated aluminium (Mazey *et al.*, 1962). McIntyre (1967) made a diffraction contrast analysis of small loops in neutron irradiated copper and found them to be interstitial in nature, in disagreement with the results of Rühle *et al.* (1965). Reviews of the subject have been given by Whelan (1963), Rühle (1969), Wilkens (1970) and by Eyre (1973). An application of the weak-beam technique to the high resolution imaging of small Frank dislocation loops has been given by Jenkins *et al.* (1973).

## 3.5  QUANTITATIVE INFORMATION OBTAINED BY ELECTRON MICROSCOPY

Having briefly reviewed a few fields of application of the electron microscope to the study of defects in metals, we now intend to conclude this account with a discussion of the quantitative uses of the technique which have found applications in the fields so far mentioned and in a variety of others as well.

### 3.5.1  Geometrical information

We emphasise here the instrumental feature mentioned in section 3.2.2, namely the facility of selected area electron diffraction, which gives crystallographic information about the specimen itself. This is a very powerful technique compared with X-ray diffraction because of the smallness of the area which can be selected and the very short exposure times required for photographic recording. It is possible to take diffraction patterns from areas straddling subgrain boundaries, as in fig. 3.16, and to work out the orientations of individual subgrains and the misorientations across sub-boundaries. Two methods are available. Firstly there is the splitting of transmission diffraction spots which enables misorientations of $\sim\frac{1}{2}°$ to be measured. Secondly there is the splitting of Kikuchi lines, in which misorientations of $\sim\frac{1}{20}°$ can be measured. The methods are discussed by Hirsch *et al.* (1965). Having determined the orientation of the specimen, it is easy to determine slip planes of dislocations and fault planes by observing the traces of these planes on the specimen surface for various sections. The thickness of the specimen may then be determined from the projected width of a slip plane. It may also be determined from the number of fringes visible in a bend or thickness extinction contour (Siems et al., 1962; Delavignette and Vook, 1963). Thus information about the volume density of defects may be obtained from the areal density on the micrograph.

An example is the study of the relation between flow stress $\tau$ and the dislocation density $\rho$. Work hardening theories (Seeger et al.,1957; Hirsch, 1964) require that $\tau = \alpha G b \sqrt{\rho}$, where $\alpha$ is a constant approximately equal to 0.5. This formula has been verified for a number of metals (Bailey and Hirsch, 1960; Carrington et al., 1960; Steeds, 1966). For further examples of measurement of the density of defects see section 3.5.4.

Reference should also be made at this point to the increasing use of the stereo method for examining the depth distribution of defects, particularly of small clusters produced by radiation damage. Normally two micrographs of the specimen are recorded, the specimen being tilted by an angle of about 5 to 10° in its plane between the two micrographs. Complications can arise if a random tilt axis is chosen, because inevitably the diffraction conditions and hence the image contrast can change markedly (i.e. $s$ changes in fig. 3.4(c)). Basinski (1962) has pointed out that this difficulty can be overcome and good stereo photographs obtained if the specimen is tilted about the normal to the reflecting planes in fig. 3.4(a), i.e. about the axis $\mathbf{g}$ in fig. 3.4(b), (c). In this situation $s$ does not change. The pair of micrographs is examined in a stereo viewer and the depth distribution can be strikingly revealed. Experienced observers claim that depth variations as small as 10 Å can be detected in this way.

### 3.5.2  Measurement of stacking fault energy†

The methods available for determining this parameter, which is very important in work hardening theory because it controls the width of extended dislocations, are somewhat limited. One of the most fruitful has been the study of extended three-fold dislocation nodes as shown in figs. 3.20 and 3.22(c). In this method (Whelan, 1959b), the stacking fault energy $\gamma$ is equated to the force per unit length acting perpendicular to the partial at $A$ in fig. 3.22(c) due to its line tension ($\frac{1}{2}Gb_p^2$) and radius of curvature ($R$). A simple treatment, neglecting the effect of other partials, gives $\gamma = Gb^2/2R$. Thus a measurement of the radius of curvature gives $\gamma$. For stainless

† See also section 2.13.

steel (18/8) a value of $\gamma \simeq 13$ erg cm$^{-2}$ was obtained. Howie and Swann (1961) and Köster et al. (1964) have measured values of $\gamma$ in this way for a number of alloys of copper and nickel; they obtained values in the range 2–20 erg cm$^{-2}$. Intrinsic and extrinsic nodes in Au–Sn alloy and segregation effects have been studied by Loretto (1964, 1965a), while stacking fault energies in silver and Ni–70wt.%Co alloy have been studied by Loretto et al. (1964) and by Loretto (1965b). Care must be taken in distinguishing between true extension of nodes and apparent extension due to excitation of two strong Bragg reflections (Booker and Brown, 1965; Shaw and Brown, 1967). Improvements in the elasticity theory treatment of the extended node configuration, taking account of the other partials, have been given by Brown (1964) and Jøssang et al. (1965).

The node method is best suited to materials where $\gamma$ is low (less than ~30 erg cm$^{-2}$) so that the width of the extended node is large compared with the width of an unextended dislocation image. For metals of higher stacking fault energy the method of annealing of faulted dislocation loops in quenched metals may be employed (Edington and Smallman, 1965; Dobson and Smallman, 1966; Dobson et al., 1967). This method is based on the climb theory of annealing of loops (Silcox and Whelan, 1960). For faulted loops of large enough radius the climb rate is controlled by the stacking fault energy, and observations of this using a high temperature stage as described in section 3.2.2 may be used to estimate $\gamma$. Values of $\gamma$ for aluminium of 280 erg cm$^{-2}$ and 135 erg cm$^{-2}$ have been reported by Edington and Smallman (1965) and by Dobson et al. (1967).

Recently stacking fault energies of pure materials and alloys have been estimated using the weak-beam technique (section 3.3.2) to reveal both the geometry of nodes and the separation of partial dislocations (ribbon width) at straight dislocations. The high resolution of the positions of dislocations obtained with this method enables separations of partial dislocations as small as 20 Å to be measured. Using isotropic elasticity theory Cockayne et al. (1969) estimated $\gamma$ for Cu–10wt.%Al alloy to be $(10.4 \pm 1.8)$ erg cm$^{-2}$ from a measured ribbon width of $(120 \pm 20)$ Å at an edge dislocation. Subsequent refinements of the technique have enabled pure materials to be examined and the interpretation to take account of anisotropy. The following are some references together with values of $\gamma$ obtained and the interested reader should refer to these for details: Ray and Cockayne (1971) (Si, ribbon width, $\gamma = (51 \pm 5)$ erg cm$^{-2}$, Si, node,

$\gamma = 50 \pm 15$ ergcm$^{-2}$); Cockayne *et al.* (1971) (Cu, ribbon width, $\gamma = (41 \pm 9)$ ergcm$^{-2}$, Ag, ribbon width, $\gamma = (16.3 \pm 1.7)$ ergcm$^{-2}$); Jenkins (1972) (Au, ribbon width, $\gamma = (32 \pm 5)$ ergcm$^{-2}$); Ray and Cockayne (1973) (Ge, ribbon width, $\gamma = (60 \pm 8)$ ergcm$^{-2}$); Ray *et al.* (1970) and Crawford *et al.* (1973) (Fe–Al alloys). These values of $\gamma$ may be compared with those obtained from measurements on nodes and other defects, prior to the development of this technique – see section 2.13, table 2.3.

### 3.5.3 Determination of Burgers vectors and the nature of faults and inclusions

*Dislocations and loops.* A most useful application of electron microscopy and diffraction contrast theory is the technique of determining Burgers vectors of dislocations, either for networks of dislocations in deformed specimens or for dislocation loops formed by clustering of point defects. The latter problem is particularly important in studies of quenching and radiation damage since a determination of the sign of the Burgers vector of a loop determines the type of point

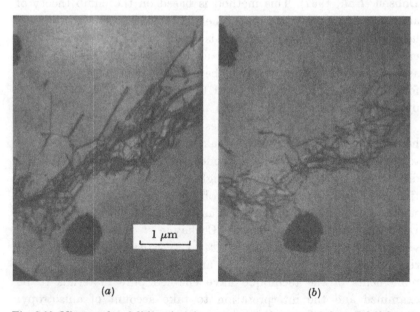

(a)                                    (b)

Fig. 3.33. Micrographs of dislocations in a copper single crystal deformed into stage II. (a) {220} reflection. Primary dislocations and secondary dislocations are visible. (b) {111} reflection. Primary dislocations are invisible; secondary dislocations are visible. (Steeds, 1966)

defect required to form it (i.e. vacancy or interstitial). The diffraction contrast principles utilised have been outlined in section 3.3.2. The direction of the Burgers vector **b** can be determined using the vanishing criterion $\mathbf{g \cdot b} = 0$. If a dislocation network is examined in several Bragg reflections consecutively by use of the goniometer stage section (3.2.2), dislocations of different Burgers vectors will vanish in different reflections (fig. 3.11), and a knowledge of these is usually sufficient to determine the direction of **b**. The magnitude of **b** may be determined from the appearance of the image profile for various $\mathbf{g}$ vectors. $|\mathbf{g \cdot b}|$ is usually 1 or 2 for whole dislocations and the profiles of the image in these cases are distinguishable and may be used to find $|\mathbf{b}|$. Details are given by Hirsch *et al.* (1965).

Figure 3.33 shows an interesting example due to Steeds (1966) for dislocation networks in copper single crystals deformed to the middle of stage II. In fig. 3.33(*a*) both primary and secondary dislocations are visible, whereas in fig. 3.33(*b*) only the secondary dislocations are strongly visible. Obviously the method is very useful for separating different dislocations in tangles. Methods of orientating the specimen quickly for a given Bragg reflection by use of Kikuchi lines have been given by Basinski (1962) and by Steeds (1966).

In order to determine the type of a dislocation loop it is necessary to determine not only the direction of the Burgers vector but also its sense (using a given convention for defining the Burgers vector). This may be done using the fact (noted in 3.3.2) that the image of a dislocation lies to one side of the centre when the foil is tilted from the exact Bragg condition so that $s \neq 0$. Thus the image of a dislocation loop lies entirely inside or entirely outside the core, depending on the sign of $(\mathbf{g \cdot b})s$ and the sense of inclination of the loop in the foil (Howie and Whelan, 1962). Experiments may be performed by keeping $\mathbf{g}$ fixed and by varying $s$, or by tilting to the opposite side of a bend contour, i.e. by keeping $s$ positive and reversing $\mathbf{g}$. The former method was used by Groves and Kelly (1961, 1962) for loops in deformed and annealed MgO crystals, and the loops were found to be vacancy in character. The latter method has been employed to study loops in $\alpha$-irradiated aluminium by Mazey *et al.* (1962); the loops were found to be interstitial in character. Small loops in neutron-irradiated copper ('black spot defects') have been considered by Rühle *et al.* (1965) and by McIntyre (1967). Feltner (1966) has found evidence for both interstitial and vacancy loops in fatigued aluminium. A simple method of determining the nature of loops from the appearance of images on tilting through an extinction contour has been given by Edmondson and Williamson (1964).

*Stacking faults and inclusions.* In the face-centred cubic lattice two types of fault can exist on {111} planes. They are known as intrinsic and extrinsic faults, and correspond to removing a close-packed layer of atoms and to inserting a layer, respectively. Both faults can also be formed by shear. The extrinsic fault, which is really a double layer fault, has a net shear which is opposite to that for an intrinsic fault. Diffraction contrast experiments can therefore be used to distinguish between them. The method employed is to take bright-field and dark-field micrographs of a fault as shown in fig. 3.15. It can be shown (Hashimoto *et al.*,1962) that, for a thick enough crystal, the contrast of the edge fringe on the bright-field image of the fault (fig. 3.15($a$)) is determined by the phase angle $\alpha = 2\pi\ \mathbf{g} \cdot \mathbf{R}$ (section 3.3.2). If the edge fringe is bright $\alpha$ is positive; the converse is true if the edge fringe is dark. Thus if $\mathbf{g}$ is known, $\mathbf{R}$ may be determined. If the sense of inclination of the fault is known, the type of fault may then be determined. The sense of inclination may be determined by examining the asymmetry of the dark-field image (fig. 3.15($b$)). Details are given by Hashimoto *et al.* (1962) and by Hirsch *et al.* (1965). Faults in Cu–Ga alloy have been studied by Art *et al.* (1963), and in Cu–Al, Cu–Ge, Ni–Co alloys by Howie and Valdrè (1963). All faults were found to be intrinsic. This might be expected since the intrinsic fault is thought to have the lower energy, although extrinsic faults have been observed in other materials (Aerts *et al.*, 1962; Loretto, 1964, 1965$a$). More complex faults in hexagonal AlN have been studied by Drum (1965) who has used vanishing conditions to determine the displacement vector at unknown prismatic faults. Stacking faults with phase angle $\pi$ were studied by Drum and Whelan (1965) for AlN and by Van Landuyt *et al.* (1964) for rutile. Rules for determining the type of fault have also been formulated by Gevers *et al.* (1963).

Small inclusions which are coherent with the matrix in metallurgical systems may also be studied by diffraction contrast techniques. For example, in the Cu–Co system small cobalt-rich spherical clusters form on ageing, and it is of interest to obtain information about misfit parameters and strain in the matrix. The problem has been studied by Phillips and Livingston (1962) using the kinematical theory and by Ashby and Brown (1963$a$, $b$) using dynamical theory.

The reader is referred to their work for further details and to Hirsch *et al.* (1965) for a review.

*Partial dislocations.* Since a partial dislocation has a Burgers vector which is not a lattice translation vector, the parameters $n = \mathbf{g} \cdot \mathbf{b}$ (section 3.3.2) occurring in diffraction contrast theory need not

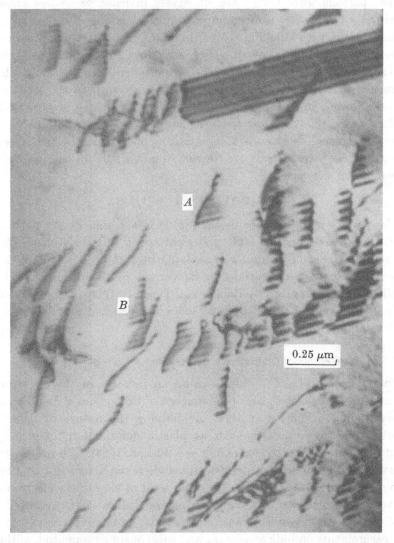

Fig. 3.34. Extended dislocations in Cu–7wt.%Al alloy. Notice that the visible partial dislocation occurs on different sides of the fault at different dislocations (e.g. *A* and *B*), showing that dislocations of opposite sign are present. (Howie, 1960*a*)

necessarily be an integer as for a whole dislocation; also, the partial dislocation must have a stacking fault on one side. For both Shockley and Frank partial dislocations in the face-centred cubic lattice, $n$ may take values 0, $\pm\frac{1}{3}$, $\pm\frac{2}{3}$, $\pm 1$ etc. Howie and Whelan (1962) have considered the images of screw Shockley partials for various values of $\mathbf{g}\cdot\mathbf{b}$ and have shown that there is an invisibility criterion such that partials with $n = \pm\frac{1}{3}$ are effectively invisible as well as those with $n = 0$. Fig. 3.34 is an example of this behaviour where extended dislocations with stacking fault fringes are visible. Only one partial bounding the fault shows up as a dark line. Silcock and Tunstall (1964) studied partial dislocations in more detail, in particular the images of Frank partials. They applied their results to the study of faults formed in stainless steel in which NbC had been precipitated. They were able to show that the faults were extrinsic and were bounded on one side by a Frank partial and on the other by a Shockley partial. The two partials are formed from a whole dislocation by a reaction of the type

$$\tfrac{1}{2}[101] \rightarrow \tfrac{1}{6}[1\bar{2}1] + \tfrac{1}{3}[111].$$

The partial with Burgers vector $\frac{1}{3}[111]$ is a sessile Frank partial, and growth occurs by NbC precipitating at this dislocation, the volume expansion being accommodated in part by advance of the partial. Further applications to faults and partial dislocations in silicon have been given by Booker and Tunstall (1966), and to double faults and partial dislocations at loops in quenched aluminium by Tunstall and Goodhew (1966).

### 3.5.4    The electrical resistivity of defects

Dislocations and point defects cause an increase in the electrical resistivity of a metal, and measurement of this quantity has been a very popular indirect method of studying the nature of defects introduced by processes such as plastic deformation, quenching and radiation damage (for review see Whelan, 1963). The resistivity is usually measured at low temperatures where the lattice contribution is minimised. It is of interest from the point of view of the electronic theory of resistivity to have experimental values of the resistivity of defects, and these are most conveniently obtained by resistivity measurements on bulk specimens and subsequent examination of the microstructure by electron microscopy. A systematic study of defects produced by quenching in gold and aluminium was made by

Cotterill (1961, 1963). In gold the resistivity can be measured before and after ageing to form the tetrahedra discussed in section 3.4.2. The size of the tetrahedra can be measured and their density can be estimated by the methods indicated in section 3.5.1. It is therefore possible to obtain the vacancy concentration if it is assumed that all the vacancies are absorbed at the tetrahedra. Thus it is possible to estimate both the resistivity $\rho_V$ of vacancies and the resistivity of stacking faults $\rho_{SF}$. In this way Cotterill (1961) obtained for gold the values $\rho_V = (2.4 \pm 0.4)\mu\Omega$ cm per atomic per cent vacancies and $\rho_{SF} = (1.8 \pm 0.3) \times 10^{-13}$ $\Omega$ cm per unit stacking fault density. The latter value is consistent with a theoretical value for copper obtained by Howie (1960b) by a pseudo-potential calculation. Similarly, the resistivity of dislocations and vacancies in aluminium may be measured by studying the resistivity of quenched specimens (Cotterill, 1963).

## 3.6 CONCLUSION

From what has been mentioned in this review it is clear that the electron microscope is a very powerful tool for the investigation of defects in crystals. Several important fields of application, such as those of semiconductors and magnetic materials, have necessarily been omitted from this account for the sake of brevity. For further information on these and additional topics the reader should consult the book by Hirsch et al. (1965). Undoubtedly, many more new problems will be studied with the technique. Comprehensive studies, including electron microscope observations taken in conjunction with the measurement of other physical properties such as mechanical properties, resistivity, and energy loss spectra, would seem to be the most fruitful fields for future investigation.

### Acknowledgment

The writer is indebted to numerous colleagues who kindly supplied photographs.

He is also extremely indebted to Professor Sir Nevill Mott FRS in whose laboratory he has obtained so much fruitful experience in the electron microscope field, for interest and encouragement over many years of work.

# CHAPTER 4

# SOLUTION AND PRECIPITATION HARDENING†

## *by* F. R. N. NABARRO‡

## 4.1 INTRODUCTION

This article is concerned with the hardening of solids, mainly metals, by various arrays of solute atoms. Random solid solutions, coherent and incoherent precipitates are considered. In addition, some attention is given to alloys containing long-range order, which have recently been studied in detail. The general theory of dislocation locking by atmospheres of solute atoms is omitted, but some recent developments are described. A study of the effect of the displacement of atoms of the matrix to produce vacant lattice sites or interstitial ions should logically form part of the present work. Largely on account of its importance in nuclear reactor technology, a great deal is known about this topic, and the reader is referred to the available books and conference reports (Dienes and Vineyard, 1957; IAEA, 1962–3; Strumane *et al.*, 1964; Chadderton, 1965).

## 4.2 TYPES OF LATTICE DEFECT

The presence of any imperfections in the lattice hinders the motion of dislocations through a crystal. Intrinsic point defects, vacant lattice sites and interstitial ions, are inevitably present in thermal equilibrium (see chapter 1). They may aggregate to form divacancies, larger aggregates, or possibly large spongy regions or voids. In face-centred cubic metals, the concentration of vacancies near the melting point may reach almost 1 in $10^3$ (Simmons and Balluffi, 1960*a*); irradiation introduces vacancies and interstitial ions in large and essentially equal numbers. In solid solutions, solute atoms may either substitute for atoms of the matrix or lie in interstitial positions.

† This work was carried out while the author was on leave at the Cavendish Laboratory, University of Cambridge, 1967.
‡ Dr Nabarro is City of Johannesburg Professor of Physics and Head of Department of Physics, University of the Witwatersrand, Johannesburg.

The pattern of elastic strain round an isolated substitutional atom generally has the symmetry of the parent lattice (although there are exceptional cases such as nitrogen in diamond), but the interstitial sites may have lower symmetry and produce strain fields of correspondingly lower symmetry. For example, carbon atoms in the body-centred cubic iron lattice lie at points displaced $a/2\langle 100\rangle$ from the iron atoms. The surroundings of these points have tetragonal symmetry, and the carbon atoms produce a tetragonal distortion of their surroundings. A substitutional atom may be bound to a vacancy, producing a strain field involving the Legendre polynomials $P_0$, $P_1$ and $P_2$.

Solute atoms may be dispersed at random. If they repel one another there will be a degree of short-range order in which the probability that a site is occupied by a solute atom is reduced if a neighbouring site is known to be occupied by a solute atom. The reverse situation, clustering, occurs if solute atoms attract one another. If the concentration of solute atoms and the repulsive forces between them are large, and the temperature is low, long-range order sets in. Domains are formed within which one set of possible sites for solute atoms becomes preferred, while another set, previously equivalent, is not favoured. A domain boundary separates this domain from one in which a different set of sites is preferred.

If solute atoms attract one another, or if a stable chemical compound can be formed between solute atoms and atoms of the solvent, a reduction in temperature may lead to the formation of a precipitate. Kelly and Nicholson (1963) have given a detailed account of the processes which occur. The first stage in the segregation of substitutional solute atoms is usually the formation of Guinier–Preston zones which have become enriched in solute atoms while retaining a one-to-one correspondence between lattice sites in the zone and lattice sites in the matrix. This is a fully coherent precipitate. If the solute and solvent atoms differ a little in radius (e.g. Ag or Zn in Al) the zones are roughly spherical and in isotropic elasticity the large components of strain around them are pure shears. If the misfit is large (e.g. Cu in Al or Be in Cu) the strain energy is reduced if the zones are thin sheets parallel to {100}. The strain field is then close to that of a prismatic dislocation lying along the edge of the disc. At a later stage the strain energy may be released by two independent mechanisms. The volume misfit between the particle of precipitate and the space it occupies in the lattice may be relieved if vacancies can move into or out of the precipitate, possibly by the formation of prismatic dislocation loops close to each precipitate particle. The precipitate lattice then remains parallel to that of the matrix, and coherent with it over most of the interface, the inequality of lattice spacings being taken up by a grid of misfit dislocations. Alternatively, the number of atoms in the precipitate may remain equal to the number of lattice sites taken from the matrix, so that there is still a volume misfit, but the one-to-one correspondence between the atomic sites in the matrix and in the precipitate is lost, so that the shear strain in the precipitate is relieved. The precipitate is then said to be incoherent. On prolonged annealing, both of these processes occur, and the precipitate particles are no longer strained or sources of strain but have a different lattice from that of the matrix and are in many cases intrinsically harder. In many alloy systems the picture is complicated by the formation of metastable intermediate phases. By varying the solute concentration and the heat treatment it is possible to control the separation and the diameter of the particles of any type of precipitate independently.

Our aim is to predict the mechanical properties of a crystal containing any of the types of imperfection which have been enumerated.

## 4.3  THE INTERACTION OF DISLOCATIONS AND DEFECTS

### 4.3.1  Long-range interactions

The most important distinction is that between long-range interactions and short-range interactions. The long-range interactions occur between the stress field of the dislocation and that of the imperfection. If the shortest distance from a long dislocation to a small imperfection is $r$, the stress which the dislocation produces at the imperfection is of order $\mu b/2\pi r$. If the volume of a spherical imperfection is $V = \frac{4}{3}\pi r_0^3$ and the strain at its surface is $\epsilon = \delta r_0/r_0$, the energy required to form it in its present position differs from the energy required to form it far from the dislocation by a quantity of order $\mu b\epsilon V/r$. We obtain the same result by imagining the dislocation to come in radially from infinity to its position distance $r$ from the imperfection. When the distance of the dislocation from the imperfection is $\rho$, the stress which the imperfection exerts on the nearest point of the dislocation is of order $\mu\epsilon V/\rho^3$. A force less than or equal to $\mu b\epsilon V/\rho^3$ per unit length acts on a length of dislocation of order $\rho$.

The total force between the dislocation and the imperfection is thus of order $\mu b \epsilon V / \rho^2$, and the work done in bringing the dislocation from $\rho = \infty$ to $\rho = r$ is of order $\mu b \epsilon V / r$. Since the force between the dislocation and the imperfection is of order $r^{-2}$ the total force between unit length of the dislocation and any similar imperfections in an element $(dr, d\theta)$ is of order $r^{-2} r \, d\theta \, dr$, and the total contribution of distant elements in a given direction may be infinite. If the interaction energy decreases more rapidly than $r^{-1}$ distant elements of the crystal contribute negligibly to the force on unit length of dislocation and the interaction is of short range. Either the long- or the short-range interaction may predominate when the separation of the dislocation and the imperfection is of order $b$. We shall first consider the effect of the long-range interaction when the separation is small.

A special interest (Fleischer, 1962$a$; Fleischer and Hibbard, 1963; Frank, 1967$a$, $c$) attaches to imperfections which produce tetragonal distortions. A defect which produces a strain field having spherical or cubical symmetry has no long-range interaction with screw dislocations. If the dislocation is dissociated the two partials may not be simultaneously in screw orientation but the interaction is still weak. If short-range interactions can be neglected a crystal containing defects of this kind should have a low yield stress because screw dislocations can initially move freely, but a high initial rate of work hardening because edge dislocations accumulate as deformation proceeds. A second effect turns out to be of greater importance. In the formula for a force between a dislocation and an impurity atom, $F \approx \mu b \epsilon V / r^2$, we may put for the volume of the atom $V \approx b^3$, and for its closest distance of approach, which occurs when the impurity atom lies immediately above or immediately below the glide plane of the dislocation, $r \approx b$. This leads to $F \approx \mu b^2 \epsilon$, while for a tetragonal strain $\epsilon$ Fleischer found a similar formula with a numerical coefficient of about 0.1. Substitutional atoms will not be present in solid solution if $\epsilon \geqslant \frac{1}{6}$ so we may take $F \approx \mu b^2 / 40$ as a reasonable estimate of the maximum force produced by a typical substitutional defect. Tetragonal defects may have $\epsilon \approx \frac{1}{2}$ or even $\epsilon \approx 1$, leading to $F \approx \mu b^2 / 10$. The reason the lattice will tolerate these large shear strains while it will tolerate only appreciably smaller dilatations is presumably the same feature of the interatomic forces which causes Poisson's ratio in close-packed metals to lie around 0.30–0.33, close to the upper end of the permissible range 0.5–(−1). The lattice resists dilatations more than it resists equal shears.

The increase in force from $\mu b^2/40$ to $\mu b^2/10$ produces a more than proportional increase in the flow stress because the obstacles to dislocation motion have become 'strong' instead of 'weak'. We suppose that the force $F$ acts over a short length $l$ of the dislocation, bending it into an arc of radius $R$ so that the dislocation turns through an angle $\theta = l/R$ in passing the obstacle. The value of $R$ is given by $R = El/F$, where $E$ is the effective line tension of the dislocation. For a bulge of radius $R$, we may take $E = (\mu b^2/4\pi)\ln(R/b)$. Since $R$ is small, this is much less than the value $\frac{1}{2}\mu b^2$ which holds for a long dislocation loop. The value of $\theta$ increases more rapidly than the value of $F$; for $F = \mu b^2/40$, $\theta \approx 5°$, while for $F = \mu b^2/10$ and $l < 4b$ Fleischer found $\theta = 180°$. In the former case the dislocation is little deviated by the obstacle. The points at which it is pinned are separated by distances which are large compared with the average separation of obstacles in the glide plane (fig. 4.8). In the latter case the dislocation changes direction sharply at each obstacle and neighbouring pinning points on a dislocation are separated by distances little greater than the average separation of obstacles in the glide plane (fig. 4.9).

### 4.3.2   Short-range interactions

Short-range interactions may also produce 'strong' obstacles. A foreign atom has elastic constants different from those of the matrix; if its rigidity $\mu$ and bulk modulus $K$ are less than those of the matrix atoms, and the foreign atom is introduced into the strained region close to a dislocation, the strain energy is reduced and the foreign atom is bound to the dislocation. Since the strain energy density at a distance $r$ from a dislocation is of order $b^2\mu/8\pi^2r^2$, the binding energy of an atom with a rigidity differing from that of the matrix by $\delta\mu$ is of order $b^5\delta\mu/8\pi^2r^2$. The exact calculation of this interaction requires some care; we shall quote the result of Saxl (1964). Saxl considered with this term another which is of the same form, namely the interaction of the dilatation associated with the foreign atom with the dilatation which occurs near to a screw dislocation as a result of non-linear elastic effects. Let the observed dependences of lattice parameter $a$, bulk modulus $K$ and rigidity $\mu$ on the atomic concentration $c$ of the foreign atoms be represented by

$$\epsilon_a = \mathrm{d}(\ln a)/\mathrm{d}c$$

$$\epsilon_K = \mathrm{d}(\ln K)/\mathrm{d}c$$

and

$$\epsilon_\mu = \mathrm{d}(\ln \mu)/\mathrm{d}c. \tag{4.1}$$

Let the foreign atom be a sphere of radius $r_0$, and lie a distance $r$ from the axis of a straight dislocation. Suppose that in the case of an edge dislocation the foreign atom lies at an angle $\theta$ above the glide plane, and that in the case of a screw dislocation the dilatation $\Delta$ (which is proportional to the square of the strain), arising from non-linear effects, is given by

$$\Delta = \mathscr{K} b^2 / 4\pi^2 r^2. \qquad (4.2)$$

Saxl estimated $\mathscr{K} \approx 1$, but Haasen (private communication) gives the more reliable estimate $\mathscr{K} \approx 0.25$ for copper, with an upper limit of $0.47$.

Also let $\epsilon_\mu K$ be a function which, when Poisson's ratio $\nu$ is $\frac{1}{3}$, becomes

$$\epsilon_\mu K = 1 - 0.3\epsilon_K/\epsilon_\mu. \qquad (4.3)$$

Then the total interaction energy $E_e^{\text{tot}}$ of an undissociated edge dislocation is given by

$$E_e^{\text{tot}} = 4\mu r_0^3 \, \epsilon_a (b/r) \sin \theta + [\mu/6\pi(1-\nu)^2] \, r_0^3 \, \epsilon_\mu (b/r)^2 \, (1 - \epsilon_\mu K \sin^2 \theta),$$
$$\qquad (4.4)$$

while the interaction energy $E_s^{\text{tot}}$ of an undissociated screw dislocation is given by

$$E_s^{\text{tot}} = -2\mathscr{K}[(1-\mu)/\pi(1-2\nu)] \, \mu r_0^3 \epsilon_a (b/r)^2 + (\mu/6\pi) \, r_0^3 \epsilon_\mu (b/r)^2. \qquad (4.5)$$

In addition to the interactions already considered, there is an electrostatic interaction between the charge distribution round an edge dislocation in a metal and the charge of a substitutional ion which has a valency different from that of the ions of the matrix. This interaction is proportional to the dilatation and is thus of long range, but is usually small. Larger effects may occur in ionic crystals. There is finally the possibility of specific chemical effects in covalent crystals (Allen, 1956). It is generally believed that the difference in valency between solute and solvent atoms has a large influence on the solution hardening of metals. Fleischer (1963, 1966) found a good correlation (fig. 4.1) between the solution hardening of copper alloys and a linear combination of $\epsilon_a$ and $\epsilon_\mu$, and claimed that valency differences, as such, had no effect. The apparent valency effect was interpreted in terms of changes in elastic moduli. We shall discuss Fleischer's correlation shortly, and note, meanwhile, that in a series of alloys there may be correlations between $\epsilon_a$, $\epsilon_\mu$ and valency difference which make a statistical analysis difficult. Table 4.1 shows some of Saxl's calculated maximum interaction energies for substitutional atoms (of cubic symmetry) in copper.

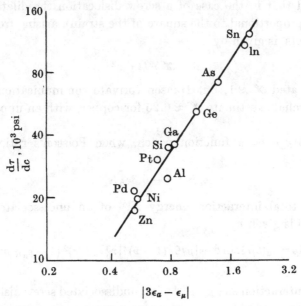

Fig. 4.1. Correlation between solution hardening parameter $d\tau/dc$ and the quantity $|3\epsilon_a - \epsilon_\mu|$. The value of $\tau$ which is used is half the flow stress of a polycrystalline specimen, adjusted to an effective grain diameter of $110\,\mu$m. It is evaluated in the range $c \approx 3$ at.%. (After Fleischer, 1963)

According to Saxl's analysis the maximum force on an edge dislocation should depend on $|33\epsilon_a - (1 - 2\epsilon_\mu K)\epsilon_\mu|$, which with $\epsilon_\mu K \approx 0.5$ becomes simply $33|\epsilon_a|$. The maximum force on a screw dislocation

Table 4.1. *Maximum interaction energies of substitutional solute atoms in copper (eV negative). (After Saxl, 1964)*

|  | Edge modulus | Edge size[b] | Edge electrostatic | Screw modulus | Screw size[b] |
|---|---|---|---|---|---|
| Al | 0.013 | 0.19 | 0.04 | 0.014 | 0.060 |
| Si | 0.026 | 0.057 | 0.08 | 0.019 | 0.018 |
| Zn[a] | 0.023 | 0.16 | 0.02 | 0.021 | 0.051 |
|  | 0.16 |  |  | 0.015 |  |
| Vacancy | 0.16 | 0.057 | 0 | 0.072 | 0.018 |

[a] Two sets of experimental observations.
[b] Haasen (private communication) suggests that these values are too large by about 50 per cent as a result of Saxl's omitting a factor of $1/\pi$ in our equation (4.2) and in the first term of our equation (4.5), and of his unduly high estimate of $\mathscr{K}$.

depends on $|24\mathcal{K}\epsilon_a - \epsilon_\mu|$, which with $\mathcal{K} = 1$ becomes $|24\epsilon_a - \epsilon_\mu|$. Fleischer found empirically a poor correlation with $|\epsilon_a|$, moderate correlation with $|\epsilon_\mu|$ and with $|16\epsilon_a - \epsilon_\mu|$, but a close correlation with $|3\epsilon_a - \epsilon_\mu|$. He believed that $|3\epsilon_a - \epsilon_\mu|$ was the parameter appropriate to the short-range interaction with screw dislocations and deduced that this was the predominant interaction in solution hardening. This would only be true if $\mathcal{K}$ had the value of 0.12, about half of Haasen's estimate for copper. It may be that, in the region of intense strain in the core of a dislocation, third order elasticity theory does not represent a significant improvement on second order theory. In niobium alloys, Harris (1966) found that the solid solution hardening correlated well with $\epsilon_a$.

In an extended dislocation in a face-centred cubic lattice the atomic arrangement in the stacking fault corresponds to hexagonal close packing. One therefore expects that those solutes which reduce the energy of the hexagonal structure relative to that of the cubic structure will concentrate in the stacking fault (Suzuki, 1952, 1957, 1962; Cottrell, 1954; Flinn, 1958; Panin et al., 1963). This is usually regarded as a specifically chemical interaction.

### 4.3.3    Interactions with lattice order

The interactions which have been considered so far occur between a dislocation and an isolated solute atom or a solute atom occupying a random position in the lattice. If the solute atoms are ordered the passage of a dislocation destroys this order (fig. 4.2). Work must be

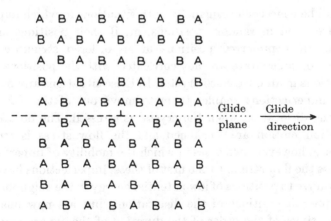

Fig. 4.2. When a dislocation moves through an ordered lattice it leaves an antiphase domain boundary (dashed line) on the glide plane

160 F. R. N. NABARRO

done to break strong A–B bonds and replaced them by weak A–A
and B–B bonds. The passage of a second dislocation restores the order
and dislocations therefore tend to travel in pairs in ordered alloys
(Koehler and Seitz, 1947; Marcinkowski *et al.*, 1961). Single dislocations
of the original lattice are partial dislocations of the superlattice. A
pre-existing domain boundary acts as an obstacle even to a pair of

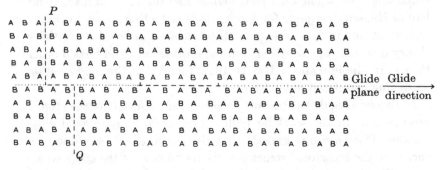

Fig. 4.3. When two dislocations travel in succession through an ordered lattice the
portion of the glide plane lying between them becomes an antiphase domain boundary.
The passage of the two dislocations also creates a ledge two atoms wide in the pre-
existing antiphase domain boundary *PQ*

dislocations because its area is increased by the passage of the dislocations (fig. 4.3). The boundary of a cluster of similar atoms or of a
particle of precipitate is an obstacle for the same reason.

## 4.4 FIXED AND MOBILE DEFECTS

We shall be concerned principally with situations in which imperfections are fixed in random positions, or, if their positions are not
random, their preferred positions in an ordered structure or in
particles of precipitate are not correlated with the positions of the
dislocations in an unstressed crystal. In these situations any mobility
of the imperfections enables them to move out of the way of an
oncoming dislocation when the crystal is stressed. The obstacles to
dislocation motion are weakened and the flow stress is reduced.
There are, however, two cases in which the mobility of imperfections
increases the flow stress. In the first of these, imperfections have been
able to move to positions of low potential energy in the neighbourhood
of the present positions of the dislocations, but are now unable to
move in times of the order of the duration of the experiment. As a
result, the dislocations are locked in their present positions and the

crystal yields with a drop of load when the dislocations escape from their atmospheres. In the second case, solute atoms jump from one site to another as a dislocation moves past them, reverting to random positions when the dislocation has passed. This causes a viscous drag on the moving dislocation and increases the stress required to cause a given rate of plastic deformation.

## 4.4.1  Dislocation locking

The basic theory of dislocation locking, and of the yield points of single crystals and polycrystals (Cottrell and Bilby, 1949; Cottrell, 1952, 1953), is too well known to need repetition here. We shall consider only some of the more recent developments of the theory (Cottrell, 1963a).

The earlier theories tended to assume that there was a stress $\sigma_0$ resisting the motion of an otherwise free dislocation and that, once the applied stress $\sigma$ exceeded $\sigma_0$, free dislocations could move and multiply in such a way as to produce the observed rate of deformation. The upper yield stress is then $\sigma_p$, the stress required to unpin a dislocation, and the lower yield stress is $\sigma_0$. This process may sometimes occur if (as in the experiments of Hutchison (1963), for example) stress concentrations are very carefully avoided, and the upper yield stress is about twice the lower yield stress. In ordinary experiments the upper yield stress is lowered by the presence of stress concentrations. Except in extreme cases, polycrystalline iron still shows an upper yield stress appreciably above the lower yield stress. We have to explain why the upper yield stress is not more readily depressed below the lower yield stress.

The explanation lies in the dependence of the velocity $v$ of a dislocation on the effective stress $\sigma - \sigma_i$ which acts on it (Johnston and Gilman, 1959; Johnston, 1962; Hahn, 1962). Here $\sigma_i$ is a frictional stress, caused perhaps by solution hardening, and opposing the motion of dislocations. While the best approximation over a wide range of $\sigma$ is probably of the form $v = v_0\exp(-\sigma_0/\sigma)$, an adequate approximation is

$$v = v_0(\sigma/\sigma^*)^n \tag{4.6}$$

where $\sigma^*$ is a constant. Here $n$ for silicon iron is about 40. The rate of plastic deformation is of order

$$\dot{\epsilon} = \rho bv \tag{4.7}$$

where $\rho$ is the density of mobile dislocations. The total density of dislocations in a piece of annealed iron is likely to be greater than $10^6$ cm$^{-2}$ but the vast majority of these are pinned by atmospheres of carbon or nitrogen and the density of mobile dislocations produced by stress concentrations may well be less than $10^3$ cm$^{-2}$. A strain of $\epsilon = 10^{-2}$ is likely to increase this density of mobile dislocations to $10^7$ cm$^{-2}$. For a given rate of plastic strain, this increase of mobile dislocation density by a factor of over $10^4$ represents a decrease of $v$ by a factor of less than $10^{-4}$. According to (4.6) $\sigma$ decreases by a factor of $10^{-0.1} = 0.8$. In this model the stress in regions of stress concentration already exceeds $\sigma_p$ when the applied stress is equal to the lower yield stress, but a stress substantially greater than the lower yield stress must initially be applied in order that the few dislocations which have been freed can move fast enough to produce the imposed rate of strain.

Within a single grain dislocations can multiply by the usual processes of bulging out between pinning points and double cross-slip. New factors arise when the slip has to spread from one grain to another. Then, if the diameter of a grain is $d$, a slip band running across a grain produces a stress $\sigma_n$ at a point in the next grain at a distance $l$ from the grain boundary, where $\sigma_n$ is given (Petch, 1953) by

$$\sigma_n = \tfrac{1}{2}(\sigma - \sigma_i)\,(d/l)^{1/2}. \tag{4.8}$$

Yield will propagate when $\sigma_n$ reaches the stress required to produce a free dislocation in the next grain. Two processes are possible: either the unpinning stress $\sigma_p$ must be obtained at a distance $l_p$ from the grain boundary, which in the average grain is the mean distance between dislocations, or the stress $\sigma_c$ required to create a new dislocation loop in the perfect crystal must be attained over the diameter $l_c$ of a critical loop. In the first case (4.8) gives for the yield stress

$$\sigma = \sigma_i + 2\sigma_p(l_p/d)^{1/2}, \tag{4.9}$$

and in the second case

$$\sigma = \sigma_i + 2\sigma_c(l_c/d)^{1/2}. \tag{4.10}$$

Typically $l_p \approx 10^{-3}$ cm, $\sigma_p$ at low temperatures is $\approx \mu/50$, while $l_c \approx 3 \times 10^{-7}$ cm and $\sigma_c \approx \mu/15$. Then $\sigma_p l_p^{1/2}/\mu \approx 1/1500$ cm$^{1/2}$, while $\sigma_c l_c^{1/2}/\mu \approx 1/30000$ cm$^{1/2}$, and at low temperatures creation is easier than unpinning. At higher temperatures, and in samples which have been only lightly aged after yielding, unpinning may be easier than creation. Since thermal activation is effective in pulling dislocations away from isolated pinning atoms, but quite ineffective in creating dislocation loops ten atoms in diameter, the yield stress should be a decreasing function of temperature when yield propagates

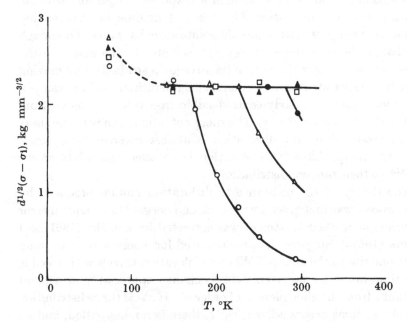

Fig. 4.4. Dependence of $d^{1/2}(\sigma - \sigma_i)$ on temperature for high purity iron containing 0.001 wt.% of carbon and nitrogen. $\bigcirc$, quenched; $\triangle$, quenched + 1 h at 140 °C; $\bullet$, quenched + 2 h at 140 °C; $\square$, quenched + 12 h at 140 °C; $\blacktriangle$, furnace cooled. Below 100 °K deformation occurred by twinning, so that the flow stress must have been higher than that recorded. This increase is not explained by the present theory of dislocation locking. (After Fisher, 1962 and Cottrell, 1963a)

by the unpinning mechanism, but independent of temperature when yield propagates by the nucleation of new dislocation loops. These predictions are verified by the observations of Fisher (1962) reproduced in fig. 4.4.

If the solution is supersaturated, dislocations may act as nuclei for incoherent precipitation (Wilsdorf and Kuhlmann-Wilsdorf, 1954, 1955; Cahn, 1967).

### 4.4.2  The drag produced by mobile defects

When a dislocation moves through a crystal the stress in a given region of the lattice changes with time, and the equilibrium distribution of solute atoms in the presence of this stress field is also time-dependent. Here again there is an important distinction between solutes which produce tetragonal strains and those which produce only strain fields of higher symmetry. In the second case the energy of interaction with the stress field of the dislocation can be appreciably altered only if the solute atom diffuses through a distance comparable with its distance from the dislocation. This takes a long time, and the moving dislocation has passed by before the solute atom has moved far enough to change the interaction energy appreciably. In the first case the solute atom can alter the axes of its tetragonal strain field by moving from its present site to an immediately neighbouring site. For example, a carbon atom in a body-centred cubic iron cell may move from $(\frac{1}{2}, 0, 0)$ to $(0, \frac{1}{2}, 0)$ or $(0, 0, \frac{1}{2})$. The energy of interaction with the shear stress produced by the dislocation is almost reversed by a single diffusive jump. After the dislocation has passed the solute atoms revert to their random distribution.

The theory of the resistance to dislocation motion produced by this process was first given by Schoeck and Seeger (1959) and an error of principle in their treatment was corrected by Eshelby (1961) and Frank (1967b). Suppose the time required for a solute atom to jump from one site to the next is $t$. When a dislocation travels with speed $v$, the time taken for its stress field to change appreciably at a point distant $r$ from the glide plane is of order $r/v$. If $r \gg vt$ the redistribution of solute atoms occurs adiabatically, there is no dissipation, and no drag on the moving dislocation. If $r \ll vt$ the dislocation has passed before any appreciable redistribution has occurred and there is again no drag produced by the redistribution of solute atoms. For a given value of $v$ the drag is contributed by solute atoms lying about $vt$ from the glide plane, which respond to the stress field but with a time lag comparable with the time in which the stress field changes appreciably. Eshelby showed that the resulting drag is proportional to $v$ when $v$ is small, but decreases as $v^{-1} \ln v$ when $v$ is large.

## 4.5  THE MEAN FORCE ON A DISLOCATION

The shear stresses in an alloy fluctuate from point to point so that the mean value of any Cartesian component of stress is zero. If a dislocation line is drawn at random in the lattice, the total force on it is just the statistical resultant of a set of forces of varying sign and is not proportional to its length. Suppose now that the ends of the dislocation line are held fixed at two widely separated points and that the dislocation has a finite line tension. Then the dislocation relaxes from its initial form into a perturbed form in which each element has reduced its potential energy. A neighbouring arbitrary configuration of the dislocation line, also passing through the two fixed points, will also relax into a configuration which is a local minimum of the energy. If the dislocation passes from one of these local minima to the other it must pass through at least one maximum of energy and it will only move under the influence of an applied force which can do work. We define the flow stress as the least stress which will allow the dislocation to propagate continuously through the alloy. If the density of mobile dislocations is low the experimentally observed flow stress may be higher than this value.

In order to calculate the flow stress we must know whether the obstacles are strong or weak (that is to say, whether a dislocation can or cannot turn through a large angle in the neighbourhood of a single obstacle) and whether they are localised or diffuse (that is to say, whether the forces they exert on a dislocation line are essentially confined to a small part of the total length of the line, or whether they act roughly equally strongly on most parts of the line). Diffuse forces were first treated qualitatively by Mott and Nabarro (Mott and Nabarro, 1940, 1948; Mott, 1952$a$; Nabarro, 1946), strong localised forces by Orowan (1948), and weak localised forces by Mott (1952), Friedel (1963$a$) and Fleischer and Hibbard (1963).

### 4.5.1   Diffuse forces

We consider an internal stress field of amplitude $\sigma_1$, having regions of constant sign of linear dimensions $\Lambda$. The force acting on unit length of dislocation is of order $\sigma_1 b$, and if we take the line tension of the dislocation to be $\frac{1}{2}\mu b^2$ it is bent into a curve of radius $\mu b/2\sigma_1$. In accordance with our definition, the force is strong when $\mu b/2\sigma_1 \ll \Lambda$ and weak when $\mu b/2\sigma_1 \gg \Lambda$.

In the case of strong forces, the dislocation lies in the troughs of the potential field in its glide plane (fig. 4.5($a$)) and individual loops of the line independently overcome individual maxima of the potential energy. The flow stress $\sigma$ is then about the mean value of the internal stress taken over a positive half cycle, so that

$$\sigma \approx 2\sigma_1/\pi. \tag{4.11}$$

The case of weak forces is less well understood. The line tension causes the dislocation to sample regions of high potential energy in the glide plane as well as regions of low energy, and the unit slip process corresponds to the advance of the dislocation on a front of width $L \gg \Lambda$ (fig. 4.5(b)). We may divide the length $L$ into $L/\Lambda$ loops, each subject to a force of order $2\sigma_1 b \Lambda/\pi$. Since these are of fluctuating sign the total force on $L/\Lambda$ loops is expected to be of order $2\sigma_1 b$

(a)

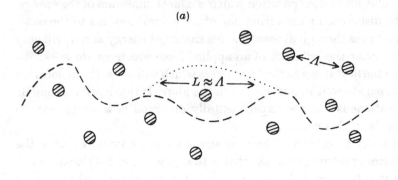

(b)

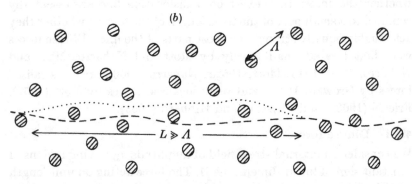

Fig. 4.5. The glide plane of a dislocation contains regions of high potential energy, centred on the shaded areas. In the elementary process of slip the dislocation moves from the dashed line to the dotted line. In (a) the obstacles are strong, and in (b) they are weak

$(L\Lambda)^{1/2}/\pi$. Equating this to the total force $\sigma b L$ produced by the applied stress $\sigma$ on the length $L$ we find

$$\sigma = 2\sigma_1(\Lambda/L)^{1/2}/\pi. \tag{4.12}$$

The difficulty is to find a realistic estimate of $L$.

We begin by noting that the angle $\theta$ through which the dislocation turns in passing a single obstacle is given by

$$\theta \approx 2\sigma_1 \Lambda/\mu b. \tag{4.13}$$

The original estimate of Mott and Nabarro was obtained by saying that in passing $n^2$ obstacles the dislocation would turn through about $n\theta$, and that the front on which the dislocation would advance was determined by writing $n\theta \approx 1$ radian. This leads to

$$L = \mu^2 b^2 / 4\sigma_i^2 \Lambda, \tag{4.14}$$

$$n^2 = \mu^2 b^2 / 4\sigma_i^2 \Lambda^2 \tag{4.15}$$

and

$$\sigma = 4\sigma_i^2 \Lambda / \pi \mu b. \tag{4.16}$$

Mott's later estimate, corresponding more closely to the behaviour suggested by fig. 4.5(b), was obtained by assuming that the distance $y$, between the dashed and the dotted lines in this curve, approximated to a parabolic function $y = \Lambda(1 - 4x^2/L^2)$. The angle $n\theta$ between the tangents to this curve at the points $x = \pm\frac{1}{2}L$ is $8\Lambda/L$. Equating these two expressions gives

$$8\Lambda/L = (L/\Lambda)^{1/2} (2\sigma_i \Lambda/\mu b)$$

which leads to

$$L = (4\mu b/\sigma_i \Lambda)^{2/3} \Lambda, \tag{4.17}$$

$$n^2 = (16\mu^2 b^2 / \sigma_i^2 \Lambda^2)^{2/3} \tag{4.18}$$

and

$$\sigma = (2\sigma_i \Lambda/\mu b)^{1/3} (\sigma_i/\pi). \tag{4.19}$$

Equation (4.14) makes $L$ a decreasing function of $\Lambda$, while equation (4.17) makes it an increasing function.

A third treatment uses second-order perturbation theory. If a dislocation line is drawn at random in the glide plane it can reduce its energy by deviating slightly under the influence of the small internal stress field. It is easy to show that the energy per unit length is reduced by

$$U = \sigma_i^2 \Lambda^2 / 8\pi^2 \mu. \tag{4.20}$$

We now suppose that a loop of dislocation of length $L$ advances into a triangular form, the ends remaining fixed and the middle advancing by $\Lambda^2/L$. The area swept out is half that which contains a single potential maximum and we believe that this corresponds to half a cycle of the fluctuation of the perturbation energy $UL$. Since this energy is composed of $L/\Lambda$ components its amplitude of fluctuation is of order $\frac{1}{2}(L\Lambda)^{1/2} U$. The line energy of the dislocation has increased by $\frac{1}{4}\mu b^2 \Lambda^4/L^3$, and an applied stress $\sigma$ does work $\frac{1}{2}\sigma b \Lambda^2$. The flow stress for this process is given by

$$\tfrac{1}{2}\sigma b \Lambda^2 = \tfrac{1}{2}(L\Lambda)^{1/2}\,U + \tfrac{1}{4}\mu b^2\,\Lambda^4/L^3. \tag{4.21}$$

This is least when

$$L = 6^{2/7}(2\pi\mu b/\sigma_1\,\Lambda)^{4/7}\,\Lambda \tag{4.22}$$

giving

$$\sigma = (6^{1/7} + 6^{-6/7})\,(\sigma_1\,\Lambda/2\pi\mu b)^{5/7}\,(\sigma_1/4\pi). \tag{4.23}$$

The results (4.22) and (4.23) are closer in form to (4.17) and (4.19) than to (4.14) and (4.16).

### 4.5.2   Localised forces

We measure the strength of a localised force by the angle $\theta$ through which it can bend a dislocation (fig. 4.6). If the forces act on the

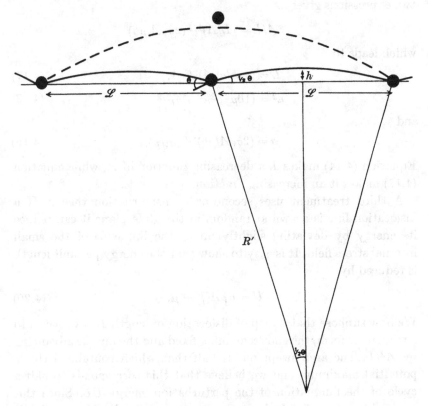

Fig. 4.6. A dislocation line which bends through an angle $\theta$ at points separated by $\mathscr{L}$ has a radius of curvature $R' = \frac{1}{2}\mathscr{L}$ cosec $\frac{1}{2}\theta$. Obstacles are represented by heavy dots

dislocation at points $\mathscr{L}$ apart, the least radius of curvature $R'$ which an applied stress can produce in the dislocation without the dislocation cutting through the obstacle is $\frac{1}{2}\mathscr{L}\operatorname{cosec}\frac{1}{2}\theta$. Assuming a line tension of $\frac{1}{2}\mu b^2$, the flow stress is given by

$$\sigma = (\mu b/\mathscr{L})\sin\tfrac{1}{2}\theta. \tag{4.24}$$

For 'impenetrable' obstacles, $\theta = \pi$ and $\mathscr{L}$ is equal to the distance $\varLambda$ between obstacles. We then obtain Orowan's formula

$$\sigma = \mu b/\varLambda. \tag{4.25}$$

Weak obstacles which attract the dislocation behave differently from weak obstacles which have a short-range repulsion for the dislocation. In the former case, a dislocation which is moving through the lattice under the flow stress zigzags from one obstacle to another. If the applied stress is removed the dislocation continues to zigzag between these obstacles. When the interactions are repulsive the dislocation has the same form under stress, but it relaxes to a smooth curve which does not touch any obstacle if the stress is removed.

The case of attractive obstacles was treated by Mott (1952a). Suppose the interaction energy between a dislocation and an obstacle is $E^{\mathrm{i}} < 0$, and suppose the mean separation of obstacles in the glide plane is $\varLambda$. Let the equilibrium form of the dislocation in the absence of stress (fig. 4.7) consist of links of length $\mathscr{L}$ which deviate by $\pm\frac{1}{2}y$ from a smooth curve. Then the energy of each link exceeds that of a straight segment which does not end on an obstacle by $E^{\mathrm{i}} + \mu b^2 y^2/4\mathscr{L}$. In moving from one configuration to a neighbouring configuration, shown dotted in fig. 4.7, the dislocation must sweep out an area

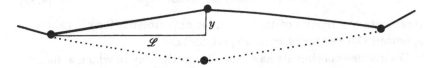

Fig. 4.7. The typical configuration of an unstressed dislocation as it runs between weak attractive obstacles. A neighbouring configuration is shown dotted

$\varLambda^2$. Hence $2\mathscr{L}y = \varLambda^2$, and the excess energy of the link is $E^{\mathrm{i}} + \mu b^2 \varLambda^4/16\mathscr{L}^3$. The energy of unit length is reduced by

$$U = -\frac{E^{\mathrm{i}}}{\mathscr{L}} - \frac{\mu b^2\,\varLambda^4}{16\mathscr{L}^4} \tag{4.26}$$

and this reduction is greatest when

$$\mathscr{L} = (-\mu b^2 \, \Lambda^4/4E^\mathrm{i})^{1/3}. \tag{4.27}$$

This value of $\mathscr{L}$ is to be substituted in (4.24).

In considering repulsive obstacles we follow Friedel (1963a) and Fleischer and Hibbard (1963). Referring to fig. 4.6, and considering only small stresses $\sigma$ for which the deviation $h$ of a dislocation segment from a straight line is much less than its length $\mathscr{L}$, we see that

$$\tfrac{1}{4}\mathscr{L}^2 = 2hR' \tag{4.28}$$

where $R'$, the radius of curvature of the segment, is given by

$$R' = \mu b/2\sigma. \tag{4.29}$$

The length $\mathscr{L}$ is determined by the requirement that the area swept out by the loop in moving to the position indicated by dashes should be the area $\Lambda^2$ of the glide plane which contains one obstacle, or

$$4\mathscr{L}h \approx \Lambda^2. \tag{4.30}$$

From (4.28), (4.29) and (4.30) we obtain the 'Friedel relation'

$$\mathscr{L} = (\mu b \Lambda^2/\sigma)^{1/3}. \tag{4.31}$$

If $\sigma$ is equal to the flow stress we may substitute this value of $\mathscr{L}$ in (4.24) and solve for $\sigma$ to obtain

$$\sigma = (\mu b/\Lambda) \sin^{3/2} \tfrac{1}{2}\theta. \tag{4.32}$$

Although (4.31) and (4.32) depend on the approximate relation (4.30) computer experiments (Foreman and Makin, 1966) show that (4.31) and (4.32) are exact for small $\theta$, while the formula

$$\sigma = (1 - \theta/5\pi)(\mu b/\Lambda) \sin^{3/2} \tfrac{1}{2}\theta \tag{4.33}$$

represents the results of the computer experiments for all values of $\theta$, within the scatter of these experiments.

Computer experiments have also shown the way in which a dislocation loop expands across the glide plane as the applied stress is increased (Kocks, 1966). The mechanism of spreading when the flow stress is reached is quite different for weak and for strong obstacles (Foreman and Makin, 1966; Kocks, 1967). When the obstacles are weak (fig. 4.8) the dislocation line retains its general direction and the line joining points which become unpinned successively lies roughly parallel to the line of the dislocation which advances on a broad front. When the obstacles are strong (fig. 4.9) the dislocation advances on a narrow front, exploiting channels along which the spacing of obstacles happens to be wider than the average.

The formulae (4.31)–(4.33) which have been derived for repulsive interactions also hold for attractive interactions when the stress

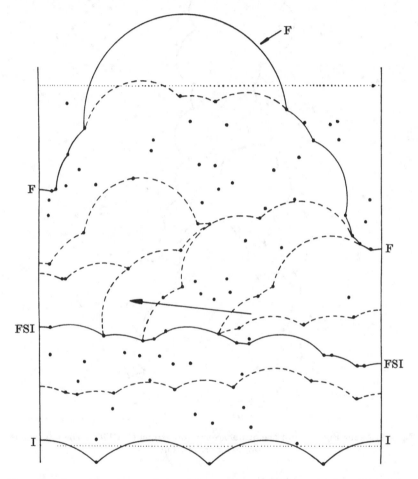

Fig. 4.8. Computed movement of a dislocation through an array of weak obstacles ($\theta = 50°$), showing initial (I) and final (F) positions. The stress was applied incrementally, and the final stress increment was applied at FSI. Broken lines show some intermediate positions and illustrate the generally transverse motion of the dislocation line (indicated by the arrow). (Foreman and Makin, 1966)

given by (4.33) exceeds that given by (4.24), that is to say, when the number of obstacles against which the dislocation is pressed by the flow stress exceeds the number with which it interacts in the absence of an applied stress.

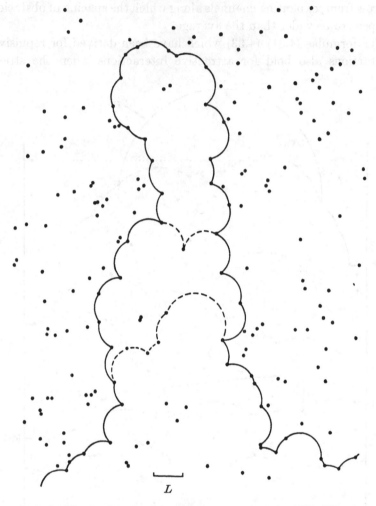

Fig. 4.9. As for Fig. 4.8, but the obstacles are now strong ($\theta = 170°$). The advance is now on a narrow front

### 4.5.3 Processions of dislocations

So far we have considered the force which acts on a single dislocation as it moves through an imperfect crystal. Suppose now that there is a procession of $n$ dislocations. The applied stress $\sigma$ exerts on the procession $n$ times the force that it exerts on a single dislocation, and the procession may move through the crystal under a smaller stress than a single dislocation. We must again consider the cases of strong and weak, diffuse and localised forces.

We shall say that diffuse sinusoidal forces $\sigma_i \cos(2\pi x/\Lambda)$ are strong if they can hold two dislocations in the same trough, and weak if they cannot. In the critical case the two dislocations will be slightly more than $\frac{1}{2}\Lambda$ apart so that the stress each exerts on the other is about $\mu b/\pi \Lambda$. The internal stress $\sigma_i$ is strong in this sense if $\sigma_i > \mu b/\pi \Lambda$, or $\mu b/\pi \sigma_i < \Lambda$. This is essentially the same as the criterion for a diffuse force to be strong (see section 4.5.1).

In the limit of very strong diffuse forces $\mu b/\Lambda \sigma_i \ll 1$ and each half-period of the internal stress contains many dislocations, provided that dislocations are freely produced by a source. We may then use the crack approximation, replacing the slip plane filled with dislocations by a freely-slipping crack subject to the same stresses. As the formal solution for one problem of this type shows (Chou and Louat, 1962), the precise value of $\mu b/\sigma_i \Lambda$ is no longer important, and the flow stress will be a moderate fraction of $\sigma_i$.

In the opposite case of very weak diffuse forces, neighbouring dislocations are separated by about $\mu b/2\pi \sigma_i \Lambda$ wavelengths of the force field. We first consider an isolated procession of $n$ dislocations. The leading dislocation is acted on by the applied stress $\sigma$ and by the repulsion $\sigma_\alpha > 0$ of all the other dislocations, and it will advance into the next potential trough when $\sigma + \sigma_\alpha = \sigma_i$. The trailing dislocation is acted on by $\sigma$ and by a retarding stress $\sigma_\omega > 0$, and it will advance only when $\sigma - \sigma_\omega = \sigma_i$. It follows that the leading dislocation will advance to infinity before the trailing dislocation moves from its original trough. The procession ultimately breaks up into a set of dislocations separated by distances much greater than $\mu b/2\pi \sigma_i \Lambda$. Although some flow occurs for $\sigma < \sigma_i$, steady flow occurs only when $\sigma = \sigma_i$. Similarly, if an 'inverted pile-up' (Bilby, Cottrell and Swinden, 1963) of dislocations is emitted by a source, the trailing dislocation will not move from the potential trough nearest to the source until the applied stress reaches $\sigma_i + \sigma_\omega$. However, this dislocation exerts a back stress of order $\mu b/2\pi \Lambda \gg \sigma_i$ on the source, which cannot operate again until the trailing dislocation has advanced by several $\Lambda$. The stress required to operate the source continuously is simply $\sigma_i$.

The case of diffuse forces with $\sigma_i \Lambda \approx \mu b$ is difficult, but since $\sigma \approx \sigma_i$ for both large and small values of $\sigma_i \Lambda/\mu b$ it is not likely that $\sigma$ and $\sigma_i$ will differ greatly for intermediate values of this parameter.

When the obstacles are strong and localised, the stress $\sigma$ required to drive a dislocation between two obstacles separated by $\Lambda$ is given by $\sigma \approx \mu b/\Lambda$. If dislocations can pile up within an obstacle-free distance $\sim\Lambda$, this stress will be lowered. The number $n$ of straight dislocations which could pile up into a distance $\Lambda$ under a stress $\mu b/\Lambda$ is, from equation (5.47) of chapter 5, $n \approx 3$. In the present case, the dislocations following the leader in the pile-up must also be curved with radii of order $\Lambda$, and we expect the co-operative motion of two or three dislocations to reduce the flow stress below that given by (4.25) by a factor of about 2.

The stress (4.32) which drives a dislocation through a set of weak localised obstacles will allow a second dislocation to approach within a distance $\Lambda/2\pi\sin^{3/2}\tfrac{1}{2}\theta$. For weak obstacles, $\theta$ is small, this distance is greater than $\Lambda$, and the following dislocation is entangled in its own set of obstacles and cannot aid the advance of the leading dislocation.

We see that there is no case in which a procession of dislocations can move continuously through an imperfect crystal under a stress much less than that required to move a single dislocation.

## 4.6  THE FLOW STRESS AT ABSOLUTE ZERO

The typical kind of observation which the theory has to explain is illustrated by fig. 4.10. When a supersaturated solid solution is aged, coherent precipitates form, grow, become incoherent, and grow again. During this process the flow stress rises and then falls again. We shall consider a solid solution containing an atomic concentration $c$ of atoms which have a radius greater than the radius of a solvent atom by a fraction $\epsilon$, and estimate the flow stress when the solute atoms have aggregated into spherical clusters of $N$ atoms. After this, we shall consider the opposite situation in which the solute atoms repel one another into a configuration of short- or long-range order.

### 4.6.1  Solid solutions

We must first decide whether the flow stress is governed by the mean amplitude of internal stress in the lattice, in which case it is given by (4.19) or (4.23), or whether it is governed by the interaction of dislocations with single atoms immediately above or below the glide plane, in which case it is given by (4.24) and (4.27) when the interaction is attractive, and by (4.32) when the interaction is repulsive.

We first interpolate between (4.19) and (4.23) by the formula

$$\sigma = (\sigma_1 \Lambda/\mu b)^{1/2}(\sigma_1/10). \tag{4.34}$$

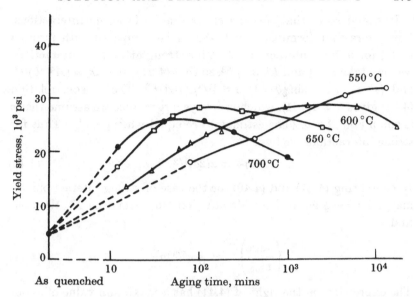

Fig. 4.10. Tensile yield stress (at 0.1 per cent plastic strain) as a function of ageing time at various temperatures for copper containing 2 wt.% Co. (After Livingston, 1959)

For the solid solution, $\varLambda$ is of the order of the distance between neighbouring solute atoms:

$$\varLambda \approx bc^{-1/3}. \tag{4.35}$$

The shear stress at a distance $r$ from a solute atom is $\mu\epsilon b^3/r^3$, and this gives

$$\sigma_1 = \int_b^{\varLambda} (\mu\epsilon b^3/r^3) r^2 \, \mathrm{d}r \Big/ \int_b^{\varLambda} r^2 \, \mathrm{d}r \approx \mu\epsilon c \ln(1/c). \tag{4.36}$$

Substituting in (4.34) we find, for the diffuse force model,

$$\sigma_{\text{diff}} \approx \tfrac{1}{10} \mu\epsilon^{3/2} c^{4/3} [\ln(1/c)]^{3/2}. \tag{4.37}$$

For the localised force model we take the two extreme cases (see section 4.3) in which $\theta = 5°$ and $\theta = 180°$, corresponding to $\sin\tfrac{1}{2}\theta \approx 0.05$ and $\sin\tfrac{1}{2}\theta = 1$. The value of $\varLambda$ is now the distance between one solute atom above or below the glide plane and the next above or below, or

$$\varLambda \approx b(2c)^{-1/2}. \tag{4.38}$$

The binding energy $E^1$ in (4.27) is, if we take $r = \tfrac{1}{2}b$,

$$E^1 \approx 2\mu b^3 \epsilon. \tag{4.39}$$

It would seem that, for the stress due to localised interactions, there were four formulae which should be compared with formula (4.37) for diffuse interactions. But for strong attractive interactions we may take $\epsilon = \frac{1}{2}$ and $E^1 \approx -\mu b^3$, so that (4.27) gives $\mathscr{L} = (\Lambda^4/4b)^{1/3}$, and (4.24), with $\sin\frac{1}{2}\theta = 1$, $\sigma \approx 4^{1/3}\mu(b/\Lambda)^{4/3}$. The Friedel relation (4.32) gives $\sigma \approx \mu(b/\Lambda)$, which is always larger since we assume that $c$ is small and $\Lambda \gg b$. (The estimates are equal when $c = \frac{1}{32}$.) Thus for strong interactions we have

$$\sigma_{\text{loc}} \approx \mu(2c)^{1/2}. \tag{4.40}$$

In comparing (4.37) and (4.40) for the case of strong interactions we may take $\epsilon = \frac{1}{2}$ as we have already done in estimating $E^1$. We then find

$$\left(\frac{\sigma_{\text{diff}}}{\sigma_{\text{loc}}}\right)^{2/3} \approx \tfrac{1}{12} c^{5/9} \ln\frac{1}{c}. \tag{4.41}$$

The expression on the right of (4.41) has a maximum value of about $\frac{1}{17}$ when $c = \exp(-\frac{9}{5}) = 0.17$. Thus $\sigma_{\text{diff}} \ll \sigma_{\text{loc}}$ and the localised interactions greatly predominate over the diffuse interactions when the interactions are strong.

When the interaction is weak (4.32) gives (with $\sin\frac{1}{2}\theta = 0.05$)

$$\sigma_{\text{loc}} \approx \mu c^{1/2}/63 \tag{4.42}$$

so that

$$\left(\frac{\sigma_{\text{diff}}}{\sigma_{\text{loc}}}\right)^{2/3} \approx 3.4\epsilon c^{5/9} \ln\frac{1}{c}. \tag{4.43}$$

The expression on the right of (4.43) has a maximum value of $2.3\epsilon$ and so $\sigma_{\text{diff}} > \sigma_{\text{loc}}$ only if $\epsilon > 1/2.3$, which is contrary to our assumption that the interaction is weak.

We conclude that the localised interactions of solute atoms immediately above and below the glide plane always predominate, and that the flow stress of a solid solution is given by

$$\sigma = \mu(2c)^{1/2} (1 - \theta/5\pi) \sin^{3/2}\tfrac{1}{2}\theta \tag{4.44}$$

where $\theta$ is the angle through which a dislocation can bend in passing an atom immediately above or below its glide plane, whether the interactions are weak or strong, attractive or repulsive. However, the derivation of (4.44) assumes that when a dislocation has 'broken away' from an obstacle its interaction with that obstacle becomes vanishingly small. If the obstacles are very weak and not very widely separated, this may not be true.

### 4.6.2  Coherent precipitates

When solute atoms assemble in spherical groups of $N$ on sites of the original lattice they produce long-range stresses. They also form obstacles which intersect the glide plane of a dislocation. Appreciable flow can occur only if dislocations can either penetrate or bow out between these obstacles. These are alternative processes and the effective process is the one which occurs under the lower stress. The flow stress is either this lower stress or the stress required to move a dislocation through the long-range stress field, whichever is the greater.

We first calculate the flow stress $\sigma$ associated with the long-range stresses. Equation (4.35) for the wavelength of the internal stress field is replaced by

$$\Lambda = b(N/c)^{1/3} \qquad (4.45)$$

while the mean stress $\sigma_1$ retains the value (4.36).

The parameter $\sigma_1 \Lambda/\mu b$, which is large for strong forces and small for weak forces, is given by

$$\sigma_1 \Lambda/\mu b = N^{1/3} \epsilon c^{2/3} \ln (1/c). \qquad (4.46)$$

Taking a large value of $\epsilon = \frac{1}{10}$, and $c = \frac{1}{100}$, makes the expression on the right about $N^{1/3}/50$; for $c = \frac{1}{10}$ it is about $N^{1/3}/20$. Thus $\sigma_1 \Lambda/\mu b$ is of order unity when $N$ is 6f order $10^4$–$10^5$, which is in the range of practical interest. Unfortunately the approximations (4.11) and (4.34) for the flow stress disagree by a factor of 6 in this range. For large $N$ we reach a flow stress given by (4.11) and equal to

$$\sigma = (2/\pi) \mu \epsilon c \ln (1/c) \qquad (4.47)$$

independent of $N$. For smaller $N$ we find from (4.34)

$$\sigma = \tfrac{1}{10} N^{1/6} \mu \epsilon^{3/2} c^{4/3} [\ln (1/c)]^{3/2}. \qquad (4.48)$$

We next suppose that the particles of precipitate form impenetrable obstacles where they intersect the glide plane. The particles are spheres of diameter about $N^{1/3}b$, each occupying a domain of linear dimensions $b(N/c)^{1/3}$. The probability that one of a set of parallel planes of area $b^2(N/c)^{2/3}$ intersects the particle is $N^{1/3}b/b(N/c)^{1/3} = c^{1/3}$, so that the area of a glide plane associated with a single penetrating obstacle is $b^2(N/c)^{2/3}/c^{1/3} = b^2 N^{2/3}/c$. The mean distance $\Lambda$ between these points is $bN^{1/3}/c^{1/2}$. Inserting this value in (4.25), we obtain

$$\sigma = N^{-1/3} \mu c^{1/2}. \qquad (4.49)$$

We suppose alternatively that the obstacles can be penetrated by the dislocations so that the particles of precipitate, instead of remaining rigid, shear homogeneously with the matrix. Suppose the particle has an ordered structure, coherent with the matrix. If the energy of unit area of antiphase boundary generated by the passage of the dislocations is $\gamma$, work $N^{1/3}b^2\gamma$ is done when the dislocation advances a distance $b$ in cutting the precipitate. The force which the particle exerts on the dislocation is therefore about $N^{1/3}b\gamma$. If the precipitate bends the dislocation through an angle $\theta$ we have

$$\mu b^2 \sin \tfrac{1}{2}\theta = N^{1/3} b\gamma$$

or

$$\sin \tfrac{1}{2}\theta = N^{1/3} \gamma/\mu b. \qquad (4.50)$$

Inserting this value of $\sin\tfrac{1}{2}\theta$, and the value $\Lambda = bN^{1/3}/c^{1/2}$ into (4.32), we obtain

$$\sigma = N^{1/6} \mu c^{1/2} (\gamma/\mu b)^{3/2}. \qquad (4.51)$$

The stresses given by (4.49) and (4.51) agree when (4.50) shows that $\theta = \pi$. Equation (4.49) is to be used when $N > (\mu b/\gamma)^3$, (4.51) when $N < (\mu b/\gamma)^3$. Assuming $\gamma$ to be about 100 erg cm$^{-2}$ and $\mu b \approx 2 \times 10^4$ erg cm$^{-2}$, it follows that particles less than 200 atoms across ($N \approx 10^7$) will be cut and the flow stress will be governed by (4.51), while large particles will not be cut and the stress will be governed by (4.49). (This discussion neglects the pairing of dislocations passing through alloys with ordered precipitates; see Kelly and Nicholson, 1971.)

We have for simplicity considered only one of the many possible short-range interactions between a dislocation and a coherent precipitate through which it cuts. Another interaction which may be important arises because the stacking-fault energy usually takes different values in the precipitate and in the matrix. The line energy of an extended dislocation therefore changes as it enters the precipitate (Hirsch and Kelly, 1965). (For discussion of various interactions see Kelly and Nicholson, 1963, 1971.)

In comparing the long-range contributions to the flow stress (4.47) and (4.48) with the short-range contributions (4.49) and (4.51) we see that each is given by expressions of different form when

$N$ is small and when $N$ is large, but that the critical values of $N$ are about $10^5$ (for the value of $\epsilon \sim \frac{1}{10}$ chosen) for the long-range contribution and $10^7$ for the short-range contribution. We shall consider only the limiting cases $N < 10^5$ and $N > 10^7$. When $N < 10^5$ the ratio $\sigma(4.48)/\sigma(4.51)$ is independent of $N$ and, for reasonable values of the parameters, of order unity. The long- and short-range stresses contribute comparably to $\sigma$ in this range. At the lower limit, $N = 1$, (4.51) is of the same form as (4.44), which governs the solid-solution hardening, though numerical agreement requires that $\sin \frac{1}{2}\theta$ in (4.44) should equal $\gamma/2^{1/3}\mu b$, which corresponds to a very weak interaction. When $N > 10^7$, the ratio $\sigma(4.47)/\sigma(4.49) \approx N^{1/3}/20 \gg 1$, so that the flow stress is determined by the long-range stresses and is given by (4.47). Under this stress, the dislocations first pass between the precipitate particles and then cut through them.

Summarising, we see that in the solid solution short-range interactions predominate. 'This roughens the slip planes' (Rosenhain, 1923), and the flow stress is given by (4.44). As the number $N$ of atoms in a particle of precipitate increases from 1 to $10^5$ the flow stress increases as $N^{1/6}$, the short-range contribution (4.51) and the long-range contribution (4.48) usually being comparable. The region of 'critical dispersion' for the chosen values of $\epsilon$, $\gamma$ lies between $N = 10^5$ and $N = 10^7$. When $N > 10^7$, the long-range stresses predominate, and the flow stress is given by (4.47) and is independent of $N$.

In many cases the coherent particles of precipitate are not spherical, but are disc-shaped. Kelly and Nicholson (1963) gave a discussion of the complicating factors which then appear.

### 4.6.3   Incoherent precipitates

A coherent disc-shaped precipitate contains shear stresses and can reduce its elastic energy without long-range diffusion by recrystallising and losing its coherence with the matrix. A spherical precipitate in an isotropic matrix is subject only to dilatation, and the matrix contains only shear stresses. The elastic energy can be reduced only by the diffusion of vacancies to or from the particle of precipitate. These vacancies may either come from distant sources, or else nucleate prismatic dislocation loops in the neighbourhood of the precipitate. In the final state, the precipitate produces no long-range stresses, but it is incoherent with the matrix. There is no longer a one-to-one correspondence between atoms in the precipitate and lattice sites in the matrix. Kelly and Nicholson (1963) suggested that if the precipitate particle is cut by the dislocation, the interface produced within the precipitate would in general have an energy comparable with that of a high-angle grain boundary, or $\mu b/30$ per unit area. Inserting this value for $\gamma$ in (4.50), we find $\sin \frac{1}{2}\theta = N^{1/3}/30$. Unless the precipitate becomes incoherent when $N^{1/3} < 30$, or unless there are special orientation relations between the precipitate and the matrix which cause the effective value of $\gamma$ to be less than $\mu b/30$, this means that incoherent precipitate particles will in general not be cut, and that the flow stress will be given by the Orowan formula (4.49). The alloy is then over-aged, and $\sigma$ is a decreasing function of $N$.

### 4.6.4  Short-range and long-range order

A. A. Griffith recognised as early as 1921 that slip by a single atomic spacing would alter the local ordering in an alloy, as illustrated in fig. 4.2, and Koehler and Seitz (1947) showed that long-range order is restored by the passage of a second dislocation across the same glide plane (fig. 4.3). The effect of order on the mechanical properties has been studied extensively and there are detailed reviews (Stoloff and Davies, 1966; Cahn, 1967).

The case of short-range order was considered quantitatively by Fisher (1953, 1954) and Cottrell (1954). Suppose that short-range ordering reduces the energy of unit area of the glide plane by $\gamma$. Then most of this energy is restored by the passage of a single dislocation and the flow stress is given by

$$\sigma = \gamma/b. \tag{4.52}$$

Further dislocations moving on the same glide plane have little effect on the ordering across the glide plane. They therefore move readily under a stress less than $\gamma/b$ and produce a pile-up which allows the leading dislocation also to move under a stress less than $\gamma/b$. The alloy therefore shows a yield point, along with deep slip lines caused by the passage of many dislocations over each of a few glide planes. Detailed calculations (Cohen and Fine, 1962) show that the passage of successive dislocations may even partially restore the short-range order, so that a dislocation is attracted towards the previous dislocation in the procession.

Local order produces local strains, both shears and dilatations. We therefore expect (Sumino, 1958) that, in an alloy in equilibrium, the state of order in the neighbourhood of a dislocation will differ from that in the matrix. The dislocation is therefore bound to the position it occupied at the time of annealing. The effect is large if the annealing temperature is just above that at which long-range ordering sets in because the state of order is then very sensitive to disturbances such as internal stresses.

When long-range order is present there is a continuous resistance to the motion of a single dislocation through the matrix, but a pair of dislocations interacts only with out-of-phase domain boundaries. The theory of the motion of paired dislocations was given by Cottrell (1954), and extended by Logie (1957) and Marcinkowski and Fisher (1963). It shows that the flow stress is greatest when the domains have a definite width $l$, typically about 12 atoms. If $\gamma$ is the energy of unit area of antiphase domain boundary, the energy of a disordered plane is $\frac{1}{2}\gamma$. Before slip the energy of unit area of the slip plane is $\alpha b\gamma/2l$, where $\alpha$ is a geometrical factor approximately equal to 6. Slip by a distance $\frac{1}{2}l$ increases this energy to $\frac{1}{2}\gamma$. The flow stress is thus given by

$$\tfrac{1}{2}l\sigma = \tfrac{1}{2}\gamma(1 - \alpha b/l)$$

or

$$\sigma = (\gamma/l)\,(1 - \alpha b/l). \tag{4.53}$$

This has a maximum

$$\sigma_{\max} = \gamma/4\alpha b \tag{4.54}$$

when

$$l = 2\alpha b. \tag{4.55}$$

While this general type of behaviour is well confirmed by experiment (Green and Brown, 1953; Biggs and Broom, 1954), Ardley (1955) showed that care must be taken to distinguish between the sizes of the domains and the degree of order within them.

Ardley and Cottrell (1953) observed a sharp maximum in the critical resolved shear stress of single crystals of $\beta$-brass at about 490°K. While the relevant elastic shear modulus shows a maximum at about the same temperature, this maximum is neither high enough nor sharp enough to explain the anomaly in the flow stress. Several explanations have been attempted. Brown (1959) pointed out that planes of atoms close to an antiphase domain boundary in thermal equilibrium would be less fully ordered than planes in the matrix, while planes of atoms close to an antiphase domain boundary pro-

duced by the passage of a dislocation would have the degree of order
of the matrix. The domain boundary between a pair of dislocations
annealed into the crystal therefore has a lower free energy than that
between two moving dislocations, and the dislocation pair is locked
to the position in which it was annealed. This effect clearly vanishes
at absolute zero, when the order is perfect everywhere except across
the antiphase boundary, and at the critical temperature, when the
long-range order vanishes, and it has a maximum between these
temperatures. Alternatively (Marcinkowski and Miller, 1961; Rudman,
1962), we notice that the passage of a dislocation through the matrix
of a partially-ordered domain affects both the long-range order and
the short-range order across the slip plane and that both contributions
to the energy must be considered. Finally (Stoloff and Davies, 1966),
we may consider that dislocations move in pairs when the long-range
order is high, but move singly and suffer a greater retarding force
when the long-range order is low.

Coherent regions of order may form within a disordered matrix.
Dislocations travel in pairs through such a structure (Gleiter and
Hornbogen, 1965). For a recent discussion of ordered alloys see Kelly
and Nicholson (1971).

## 4.7 THE TEMPERATURE DEPENDENCE OF THE FLOW STRESS

Change of temperature influences the flow stress in three ways.
Firstly, thermal vibrations may activate dislocations over fixed barriers formed by arrays of solute atoms. The theory leads to a number
of classical activation energy problems which have been surveyed
by Nabarro (1967). Secondly, individual solute atoms may move in the
neighbourhood of the dislocation under the combined influence of the
applied stress and of thermal vibrations. Finally, the whole pattern
of solute atoms may alter towards the configuration which is in
equilibrium at the new temperature. This is important in practice,
but raises no new theoretical problems.

We will first consider the situation in which solute atoms have
segregated close to dislocations. If the temperature is low enough
for appreciable segregation of solute atoms to occur on stacking
faults, the energy required to pull a loop of dislocation away from the
region of segregation is many times the thermal energy $kT$, and the
flow stress should depend on temperature only in proportion to the

elastic constants. The calculation of the thermally activated escape of a dislocation from a Cottrell atmosphere 'is a difficult and somewhat controversial topic' (Friedel, 1963$b$). The most important question is whether the large temperature dependence of the flow stress of body-centred cubic metals is caused by the interaction of dislocations with interstitial solute atoms or by the Peierls force of the perfect lattice. There is much evidence that the very large increase in the flow stress at very low temperatures is associated with the Peierls force (Conrad, 1963; Christian and Masters, 1964; Guyot and Dorn, 1967). However, Stein and Low (1966), Fleischer (1967) and Frank (1967$d$) have assembled evidence that this increase, as well as the less rapid increase observed at intermediate temperatures ($\approx 100\,^\circ$K), is associated with the presence of solutes. Cottrell (1963$b$) pointed out that these two mechanisms may co-operate to give an enhanced flow stress. Flanagan (1967) observed that the proportional limit, which depends on the displacement of dislocations from their relaxed positions into Peierls troughs, should be sensitive to the presence of impurities, while the flow stress may depend on the excitation of dislocations from Peierls troughs and be less dependent on impurities.

The temperature dependence of the flow stress in solid solutions and in alloys containing coherent precipitates which harden by their long-range stresses was calculated by Mott and Nabarro (1948). As has been emphasised by many authors (e.g. Kelly and Nicholson, 1963), the theory of the activation of a dislocation over a smooth-topped potential hill of any form always gives the same form of relation between the flow stress $\sigma(T)$ at temperature $T$ and the flow stress $\sigma(0)$ at temperature zero, provided that the temperature variation of the elastic constants is neglected. If a stress $\sigma = [1 - \eta]\sigma(0)$ is applied, the distance of the dislocation from the crest of the potential hill is proportional to $\eta^{1/2}$ and the activation energy is proportional to $\eta^{3/2}$. Flow will occur at a measurable rate when this activation energy is a constant multiple of $kT$. The flow stress is thus given by

$$\sigma(T) = [1 - \eta(T)]\,\sigma(0)$$

where

$$[\eta(T)]^{3/2} \propto T.$$

Thus

$$\sigma(T) = \sigma(0) - A T^{2/3} \tag{4.56}$$

where $A$ is a constant.

Fleischer (1962b), using an algebraic approximation to a calculated interaction function, obtained the relation

$$[\sigma(T)/\sigma(0)]^{1/2} = 1 - (T/T_0)^{1/2} \tag{4.57}$$

where $T_0$ is a constant. This relation agrees well with a number of experiments over a wide range of $\eta$ even though, for small $\eta$, (4.56) must hold and (4.57) must fail.

Phillips (1965) observed an interesting case in which the flow stress of a very simple system was an increasing function of temperature. Cobalt forms a coherent precipitate of spherical particles of face-centred cubic structure in a copper matrix. For particle sizes between 25 Å and the size which gives maximum hardening (70–140 Å), the flow stress increases with increasing temperature because the misfit $\epsilon$ of equation (4.36) increases with increasing temperature.

The activation energy is much smaller when the flow stress is determined by the cutting of coherent precipitates (Byrne et al., 1961) than it is when the flow stress is determined by long-range internal stresses. The precipitate may consist of a sheet of solute atoms only one plane thick, and the dislocation breaks the bonds between like atoms one by one, with a small activation energy. When hardening is due to long-range stresses an equal stress has to act over a segment of dislocation many atoms long, which may advance through many atomic spacings, with a large activation energy.

In the Orowan model, in which dislocations bow out between widely spaced incoherent particles of precipitate, the activation energy is much greater than $kT$ unless $\eta$ is very small, and the temperature dependence of the flow stress should arise only from the temperature dependence of the elastic constants. Unless great care is taken in experiments the flow stress is in fact governed by a dislocation forest and not by the bowing-out mechanism and a large temperature dependence is observed (Dew-Hughes, 1960).

The temperature dependence of the flow stress of crystals showing long-range order is very complicated. For example, the flow stress of $Ni_3Al$ increases with increasing temperature from 80 to 900°K (Stoloff and Davies, 1966), AgMg shows an anomaly confined to the range 380–580°K (Wood and Westbrook, 1962), and the flow stress may either rise or fall when long-range order sets in (Westbrook, 1965). Reviews are given by Westbrook (1965) and by Stoloff and Davies (1966).

## 4.8  THE STRESS–STRAIN CURVE

Alloying usually lowers the stacking-fault energy of a metal, and this makes cross-slip more difficult. In addition, the destruction of short-range order or clustering by the passage of one dislocation across a glide plane reduces the stress required to pass other dislocations across the same plane so that deep sharp slip lines are formed, containing processions of dislocations (Wilsdorf and Fourie, 1956), and deformation continues on the primary system even if the crystal rotates so that another system is more highly stressed. The observations are reviewed by Haasen (1965). There is always a long stage I of low hardening, but the deformation in this stage may occur either homogeneously, as in Ni–Co (Pfeiffer and Seeger, 1962), or by the propagation of a Piobert–Lüders band, as in $\alpha$-brass (Piercy et al., 1955; Mitchell and Thornton, 1963). The slope of stage II is independent of solute concentration over a wide range. The onset of stage III occurs only at high stresses, presumably because the low stacking-fault energy makes cross-slip difficult, and the alternative method of stress relief by dislocation climb may occur.

When coherent precipitates are formed the initial flow stress may be raised greatly and stage I is absent, but there is a long, nearly linear stress–strain curve with a slope very close to that of stage II in the pure base metal or in the supersaturated homogeneous solid solution. The slip lines are wavy, but approximate closely to those produced by slip on a single system.

When the precipitate becomes incoherent the flow stress drops to about 0.6 of its maximum value, but the rate of hardening at 2 per cent strain increases by a factor of about 5 (Byrne et al., 1961; Matsuura and Koda, 1963). The original theory of work hardening in such a dispersion-hardened alloy was due to Fisher et al. (1953). Slip is assumed to occur on a single system. Dislocations, impeded by precipitate particles, pass between them, leaving rings of dislocation round each particle. The hardening is produced by the back stress from these loops; it is at first proportional to the cube root of the plastic strain, increasing so rapidly that it makes an appreciable contribution to the stress even at a strain of 0.1 per cent (Hart, 1972). The back stress saturates when the particles of precipitate begin to shear. This mechanism of hardening should obviously be almost independent of temperature. In fact, multiple glide occurs, the rate of work hardening depends strongly on temperature (Byrne *et al.*, 1961), and the activation process for slip appears to be the same as that in the homogeneous matrix (Mitchell *et al.*, 1963). The explanation (Ashby, 1966) is that the matrix, even allowing for some residual solution hardening, is much softer than the particles of precipitate. Since there must be continuity of stress between the precipitate and the matrix, the stress concentrations at the precipitate particles which are associated with the dislocation loops in the model of Fisher *et al.* induce secondary slip in the matrix. As is observed in the electron microscope, dislocation tangles form around the precipitate particles, and the flow stress is the stress required to force glide dislocations through these tangles. There are then three stages in the work-hardening curve. Below about 2 per cent strain, a model of the type of Fisher *et al.* applies, and a large Bauschinger effect is expected and observed. Then Ashby's model becomes effective, with an increase of flow stress $\delta\sigma$ given by

$$\delta\sigma \approx 0.2\mu(bf\epsilon/d)^{1/2}. \tag{4.58}$$

Here $\mu$ is the shear modulus of the matrix, $b$ is the Burgers vector, $f$ is the volume fraction of the precipitate and $d$ the diameter of the particles, and $\epsilon$ is the strain. In the third stage, the matrix breaks away from the precipitate and the rate of work hardening is reduced by a factor of $1/\sqrt{2}$. For further discussion see chapter 5.

We finally consider the work hardening of alloys having long-range order. Work hardening is normally more rapid in the ordered alloy than in the disordered. In a strongly ordered alloy, such as one approximating to the composition AgMg, over 30 per cent of the work of deformation may remain stored in the lattice (Robinson and Bever, 1965), as compared with less than 10 per cent for a pure metal. There is little doubt that this high rate of energy absorption is caused by progressive disordering. If dislocations travelled through the disordered lattice in pairs on the same glide plane, disorder would be created only at existing domain boundaries (fig. 4.3) and the presence of very large domains would have very little influence on either the initial flow stress or the rate of work hardening. In fact it seems that large domains have little effect on the initial flow stress but contribute appreciably to the rate of work hardening. We therefore assume that a pair of dislocations may move on two glide planes which are slightly separated rather than on the same glide plane. Then the trailing dislocation creates a new antiphase boundary as it moves through the domain, instead of healing the boundary produced by the leading dislocation. Two types of mechanism have been proposed for this separation. In the first (Vidoz and Brown, 1962; Vasil'yev and Orlov, 1963), the intersection of two dislocation pairs leads to the formation of two pairs of jogs in each dislocation pair. As a result, the glide of one jogged dislocation pair leads to the formation of a tubular antiphase domain. This process should be independent of temperature and of strain rate at temperatures at which diffusion is negligibly slow. However, Kear (1964, 1966) and Davies and Stoloff (1965) have shown that the rate of work hardening of ordered $Cu_3Au$ roughly doubles on raising the temperature from 77 to 298°K. They therefore favour variants of the mechanism originally discussed by Flinn (1960), which operates in the following way. In the $Cu_3Au$ structure the energy of an antiphase domain boundary is extremely low when it lies on {100} because there are then no wrong first-neighbour bonds, whereas it is high on the glide plane {111}. There is thus a tendency for one dislocation of a gliding pair to undergo thermally activated cross-slip so that the plane in which the pairs lie is {100}, when each continues to glide in a {111} plane. This process, which obviously leads to work hardening, will occur more readily at high temperatures. The rate of hardening falls again when diffusion becomes fast enough to restore order in the wake of the moving dislocations. This mechanism should not operate in the $\beta$-brass structure which has no preferred planes on which antiphase domain boundaries have very low energies.

Ordering in body-centred cubic alloys based on FeCo, with the formation of a $\beta$-brass structure, causes the slip system to change from pencil glide parallel to $\langle 111 \rangle$ to planar glide on $\{01\bar{1}\}\langle 111 \rangle$. Consequently, the propagation of slip from one grain to another in a polycrystal is rendered more difficult. The dependence of flow stress on grain size $l$ is enhanced, $k$ in the formula $\sigma = \sigma_0 + kl^{-1/2}$ increasing by a factor of 3 (Marcinkowski and Fisher, 1965). In an alloy containing 2at.% vanadium, ordering raises the ductile–brittle transition temperature from −100°C to 450°C by the same mechanism (Johnston et al., 1965).

## POSTSCRIPT

Since this article was written, considerable advances in the statistical problem of hardening, probably representing a complete solution in some cases, have been made by R. Labusch. The first of his series of papers (Labusch, 1969) develops the mathematical formalism in the somewhat different context of the resistance to the motion of the lattice of flux lines in a type II superconductor. The same theory is then (Labusch,1970) applied to the problem of the mechanical strength of a solid solution, and a derivation of the result by a more elementary method is given in an appendix. There is a critical discussion of this and other theories in Labusch (1972); the theory is extended to treat thermally activated flow in Labusch *et al.* (1975). The general position of the statistical problem of hardening was summarised by Nabarro (1972). Lowell (1972) gives a very clear account of the physical principles underlying the application of the theory to the case of the pinning of flux lines.

## Acknowledgment

I am grateful to Drs R. L. Fleischer, E. Hart and A. Kelly, and Professor P. Haasen, for their comments on the manuscript.

# CHAPTER 5

# WORK HARDENING

## *by* P. B. HIRSCH †

## 5.1 INTRODUCTION

Since the original work due to G. I. Taylor (1934) numerous attempts
have been made to explain the work hardening phenomena in terms
of dislocation mechanisms. Much progress has been made on the
experimental side in determining the nature of the stress–strain
curves of various materials as a function of various parameters such
as temperature, strain rate, grain size, crystal orientation (in the case
of single crystals), alloy composition etc. The development of methods
for the direct observation of dislocations, mainly by transmission
electron microscopy and etch-pitting techniques, has made it possible
to study in detail the internal distribution of dislocations as a function
of deformation, and the dislocation arrangement is now quite well
established in a number of cases. Studies of slip lines have also yielded
valuable information on the scale on which the slip processes take
place. Dislocation theory has seen much development and a number
of mechanisms and dislocation interactions important in work
hardening have been established. However, in spite of a spate of
theories during the last few years designed to explain the work
hardening of single crystals, the phenomena are still not well understood. The aim of any work hardening theory is to explain the stress–
strain curve, and its dependence on temperature, strain-rate, etc.
This involves usually the assumption of a model of the dislocated
state, which is characterised by a flow stress which depends on one or
more parameters of the dislocation distribution, and the variation
of these parameters with strain. Most of the attempts to explain work
hardening phenomena have been concerned with establishing the
nature of the mechanisms controlling the flow stress; few have
tackled the more difficult problem of the variation of the dislocation
distribution with strain. The difficulty of this latter problem has been

† P. B. Hirsch is Isaac Wolfson Professor of Metallurgy at the Department of Metallurgy and Science of Materials, University of Oxford.

apparent since the original theory of Taylor (1934), who assumed the dislocations to be arranged in a regular lattice of positive and negative dislocations. The flow stress for such an arrangement was calculated, but the mechanism for the formation of this lattice, or for the decrease in lattice parameter with increasing strain, was not clear.

In this chapter we shall confine ourselves to the deformation of single crystals; we shall emphasise certain aspects which arise from work more recent than that covered in the excellent reviews in this field by Nabarro, Basinski and Holt (1964), Boček (1963) and Mitchell (1964). The subject is still one of considerable controversy, and the various points of view are presented in papers published in the reports of conferences held in 1966 in Ottawa and Chicago respectively (for proceedings see Basinski and Weinberg, 1967 and Hirth and Weertman, 1968).

## 5.2  NATURE OF STRESS–STRAIN CURVES

It has been established for some time that the stress–strain curves for face-centred cubic metals and alloys consist of three stages (fig. 5.1) (for review see Seeger, 1957; Clarebrough and Hargreaves, 1959; Mitchell, 1964; Nabarro, Basinski and Holt, 1964). Stage I is approximately linear, and the hardening rate, $\theta_I$, is low and of the same order as that for hexagonal metals, i.e. $\theta_I/G \sim 10^{-4}$, where $G$ is the shear modulus. Stage II is also linear, and the hardening rate $\theta_{II}/G \sim 1/300$; this parameter varies only over relatively small limits from one metal or alloy to another (say a factor of 2 either way). $\theta_I$ varies with crystal orientation, by about a factor of 10 over the orientations in the stereographic triangle, reaching maxima near the corners and edges of the triangle. $\theta_{II}$ on the other hand varies only by about a factor of 2 with orientation, being again greater near the corners than in the middle of the triangle. The extent of stage I (easy glide) also depends in a systematic way on crystal orientation, temperature and alloy composition; these data are reviewed by Mitchell (1964). In stage III the work hardening rate decreases again with increasing strain, and the stress at the onset of stage III increases with decreasing temperature.

During the last few years it has become clear that the three-stage hardening curve of single crystals is typical also of b.c.c. metals of high purity, over a certain range of temperatures and strain rates (Nb – Mitchell, Foxall and Hirsch, 1963; Taylor and Christian, 1965;

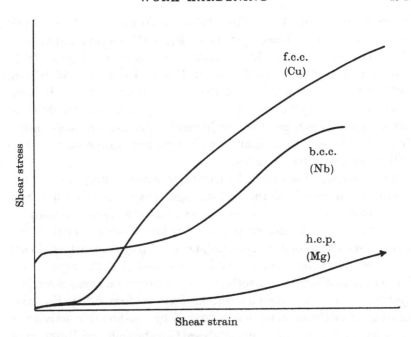

f.c.c.
(Cu)

b.c.c.
(Nb)

h.c.p.
(Mg)

Shear stress

Shear strain

Fig. 5.1. Stress–strain curves of single crystals of typical f.c.c., b.c.c. and h.c.p. metals. (Hirsch and Mitchell, 1967.) Reproduced by permission of the National Research Council of Canada from *The Canadian Journal of Physics* **45**, 663–706, 1967

Duesbery, Foxall and Hirsch, 1966; Bowen, Christian and Taylor, 1967; Foxall, Duesbery and Hirsch, 1967. Ta – Mitchell and Spitzig, 1965. Fe – Keh, 1965; Keh and Nakada, 1968), of h.c.p. metals (Zn – Boček and Kaska, 1964. Cd – Boček, Lukáč, Smola and Švábová, 1964. Mg – Boček, Hötzsch and Simmen, 1964. Co – Seeger, Kronmüller, Boser and Rapp, 1963), of Ge and In–Sb (Schäfer, Alexander and Haasen, 1964) and of alkali halides (NaCl – Hesse, 1965; Davidge and Pratt, 1964). Fig. 5.1 shows stress–strain curves for Cu, Nb and Mg at room temperature; the curves for Nb are similar to those from f.c.c. alloys. It should be noted that for h.c.p. metals stage I is generally much longer than for the other crystal structures, and stage II may not be fully developed before fracture occurs.

It is clear, therefore, that the 3-stage hardening curve is a rather general phenomena. Electron microscope observations have been made of the dislocation distributions in deformed crystals in all three stages of the hardening curve of various materials (Cu – Hirsch, 1963; Basinski, 1964; Essmann, 1963, 1965*a*, *b*; Steeds, 1966; Ag – Moon and Robinson, 1967. Au – Ramsteiner, 1966, 1967. Cu–Al alloys – Steeds and Hazzledine, 1964. Ni–Co alloys – Mader, 1963; Mader, Seeger and Thieringer, 1963. Mg – Hirsch and Lally, 1965. Co – Thieringer, 1964. Nb – Bowen, Christian and Taylor, 1967; Taylor and Christian, 1965, 1967a; Foxall, Duesbery and Hirsch, 1967. Ge – Alexander and Mader, 1962; Alexander and Haasen, 1967) and the similarities of the distributions in several different crystal systems suggests that the hardening mechanisms are in many (but not all) respects similar; the electron microscope observations will be discussed in section 5.4.

The similarities between the hardening rates in stage I of a face-centred cubic metal and the (rather longer) easy glide range in hexagonal metals led Mott (1952b) to suggest that both are due to the same mechanism: that hardening in stage I is low because very little slip takes place on 'secondary systems', i.e. on slip systems which do not bear the maximum resolved shear stress: and that the rapid linear hardening in stage II is associated with activity on these secondary systems. In fact slip line observations have confirmed that the onset of stage II of the deformation is marked by considerable activity on secondary systems (for f.c.c. metals see Clarebrough and Hargreaves, 1959; Mitchell, 1964; for b.c.c. metals see Duesbery, Foxall and Hirsch, 1966; Duesbery and Foxall, 1969). Furthermore, electron microscope observations (to be described later) show clearly that in stage I most of the dislocations belong to the primary system, while in stage II the densities of primaries and secondaries are comparable. The amount of plastic strain due to slip on secondary systems is still somewhat in doubt. X-ray diffraction determinations of the rotation of the tensile axis of the crystal as a function of strain are consistent with the view that most of the elongation is due to slip on the primary system, the contributions from secondary slip accounting only for a few per cent (Ahlers and Haasen, 1962; Mitchell and Thornton, 1964). Recently, however, careful measurements of shape changes of crystals of copper deformed in compression in stage II suggest that the plastic strain on all the secondary systems amounts to 35–50 per cent of the total plastic strain (Johnson, Kocks and Chalmers, 1968). The fact that stage II is associated with secondary slip explains why in some h.c.p. metals such as magnesium (Hirsch and Lally, 1965; Boček, 1963; Nabarro, Basinski and Holt, 1964) the easy glide stage extends over a much greater range of strain than in f.c.c. metals, since (at room temperature and below) slip on non-basal planes in magnesium occurs with much greater difficulty than slip on the basal plane.

The onset of stage III in f.c.c. metals is accompanied by the appearance of cross-slip traces joining primary slip lines (Mader, 1957; Seeger et al., 1957); this observation had led to the idea that a dynamic recovery process takes place in which screw dislocations of opposite sign on primary slip planes annihilate each other by cross-slip. While this process seems well established in f.c.c. metals and alloys, it is not clear whether a similar process marks the onset of stage III in b.c.c. metals; there is some electron microscope evidence to suggest that the density of screws in niobium is less in stage III than in stage II (Foxall, Duesbery and Hirsch, 1967), which would be consistent with a dynamic recovery process involving annihilation of screws by cross-slip. (It should be emphasised, however, that there is ample evidence from electron microscopy that cross-slip occurs on a fine scale even in stage I – see section 5.4.)

Summarising this section so far, we note that the 3-stage work hardening curve is a general phenomenon characteristic of a number of different crystal structures, that stage II differs from stage I by the considerable increase in activity on secondary systems, and that in stage III a dynamic recovery process occurs, which consists of the annihilation of screws by cross-slip.

There is one other important class of materials we shall consider. These are alloys containing impenetrable particles, for example copper with silica, beryllia or alumina particles produced by internal oxidation of Cu–Si, Cu–Be or Cu–Al alloys (Ashby, 1966; Ebeling and Ashby, 1966; Humphreys and Martin, 1967; Jones and Kelly, 1968; Jones, 1969; for recent references see Ashby, 1971). Alloys containing small volume fractions ($f$) of particles also have 3-stage stress–strain curves; this is clear from fig. 5.2 which shows a typical set of stress–strain curves from alloys with two different volume fractions, compared with the stress–strain curve of pure copper. We note that for small $f$ the hardening rate in stage II is similar to that for pure copper, and the main difference between the curves is the increase in yield stress and the fact that stage I in these alloys is approximately parabolic, with high initial hardening rates. Slip line and electron microscope observations show that in stage I most of the dislocations belong to the primary system, while in stage II the primary and secondary dislocation densities are comparable and the dislocation arrangement is similar to that found in pure Cu (Humphreys and Martin, 1967). In these alloys the particles act as well-defined obstacles to the dislocation, and the dislocation–particle

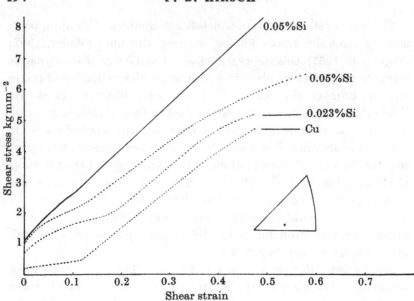

Fig. 5.2. Stress–strain curves of single crystals of copper containing silica particles (volume fractions $f$, $1.8 \times 10^{-3}$ and $4 \times 10^{-3}$ respectively), produced by internal oxidation of Cu–Si alloys. ---- 293 °K, ——77 °K. (Humphreys and Martin, 1967)

interaction leads to systematic generation of dislocation loops (Ashby, 1966; Humphreys and Martin, 1967; Hirsch and Humphreys, 1969). By varying the numbers and sizes of particles it is possible to test work hardening theories in a more exhaustive way than is possible for pure metals or solid solution alloys (see section 5.6). It should also be mentioned that crystals containing penetrable particles exhibit stress–strain curves rather similar to those from solid solutions (Kelly and Nicholson, 1963).

## 5.3 SLIP LINE OBSERVATIONS

Electron microscope replicas of slip lines in f.c.c. and h.c.p. single crystals show that in stage I the slip lines are long (e.g. in Cu typically $\sim 600$ $\mu$m) and their length can be of the order of the diameter of the crystal. In the case of copper and Ni–Co crystals the slip line lengths and separations are approximately constant in stage I, but the heights of the slip lines increase with increasing strain (Mader, 1957; Seeger, Kronmüller, Mader and Träuble, 1961; Kronmüller, 1959; Mader, 1963). The number of dislocations per slip line increases to

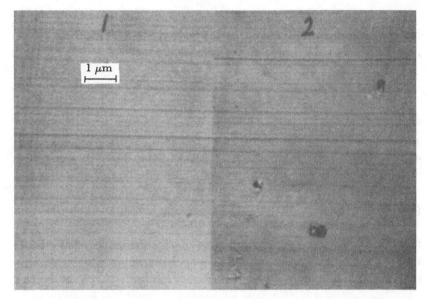

Fig. 5.3. Electron micrographs of replicas of fine slip lines on single crystals of copper deformed at 90 °K in stage I. The numbers 1 and 2 denote two successive strain intervals with total shear strains 0.048 and 0.0753. (Seeger, Kronmüller, Mader and Träuble, 1961)

~20–30 at the end of stage I, and remains constant in stage II; fig. 5.3 shows the fine slip lines in stage I in Cu (Mader, 1957). On the other hand, in Cu–20wt.%Zn crystals the slip line heights are considerably greater; according to Fourie and Wilsdorf (1959) the average number of dislocations per slip line is ~100. More recently, detailed studies of the initial stages of the deformation in Cu–Al alloys (Mitchell, Chevrier, Hockey and Monaghan, 1967) have shown that slip takes place in bands traversing the whole specimen, and that these contain slip lines containing several hundred dislocations. Transmission electron microscope observations of Cu–Al alloys also indicate that the number of dislocations per slip line can be very large (Pande and Hazzledine, 1971a).

In zinc (Seeger and Träuble, 1960) and cobalt at 90°K (Seeger, Kronmüller, Boser and Rapp, 1963) the slip line lengths and separations are found to be constant (the latter after an initial critical strain of 15–25 per cent); in magnesium at room temperature the slip line spacing decreases with increasing strain but the slip line length is constant; each slip line in an increment test contains ~70 dislocations (Hirsch and Lally, 1965).

The work of the Stuttgart group on copper and on Ni–Co alloys has shown that the lengths of the slip lines in stage II decreases with increasing strain according to the law

$$L_2 = \frac{\Lambda_2}{\epsilon - \epsilon^*} \qquad (5.1)$$

where $L_2$ is the slip distance of screw dislocations, $\Lambda_2$ a constant and $\epsilon^*$ the strain corresponding to the end of stage I. With increasing strain new slip lines appear, and these contain about 20–30 dislocations (independent of $\epsilon$) and are fully formed in a small stress interval (Mader, 1957; Seeger, Mader and Kronmüller, 1963).

In the case of b.c.c. metals the slip lines are 'wavy' due to cross-slip and because of this no measurements of slip line lengths have been made. Furthermore, the individual slip lines appear to be so fine that replicas cannot resolve them. Fig. 5.4 shows an optical micro-

500 μm

Fig. 5.4. Optical micrograph of wavy slip lines on single crystals of niobium deformed into stage I. (Foxall, Duesbery and Hirsch, 1967.) Reproduced by permission of the National Research Council of Canada from *The Canadian Journal of Physics* **45**, 607–29, 1967

graph of wavy slip lines in niobium (Foxall, Duesbery and Hirsch, 1967).

It should be mentioned that slip line measurements suffer from the disadvantage that they are essentially surface observations and describe the slip pattern at the surface. This may differ from that occurring in the interior of the crystal, but work hardening theories usually assume that the observed slip lines describe adequately the behaviour inside the crystal. There is inevitably some doubt about this. Experiments by Fourie (1967) on single crystals of copper have shown clearly that the deformation of the surface layer is considerably different from that of the interior of the crystal, and that the slip line pattern in his experiments does not reflect bulk behaviour. In dispersion-hardened alloys the slip line pattern has been found to be quite inconsistent with the slip line distribution deduced from transmission microscopy of the interior of the crystal (Hirsch and Humphreys, 1970a).

## 5.4  DISLOCATION DISTRIBUTIONS

Transmission electron microscope studies have been carried out on a number of metals and alloys (for reference see Hirsch and Mitchell, 1967; Seeger, 1968, also Ottawa and Chicago conference reports; also references in section 5.2). The structures in copper appear to be typical of those in a pure metal, those in Cu–Al alloys typical of solid solutions. The observations on copper (Hirsch, 1963; Essmann, 1963, 1965a, b; Basinski, 1964; Steeds and Hazzledine, 1964; Steeds, 1966) show that in stage I bands of primary edge dipoles are formed (fig. 5.5); at the beginning of stage II these are linked by secondary dislocations to form continuous barriers of dislocations (fig. 5.6). The structure characteristic of stage II consists of 'carpets' or 'mats' of dislocations more or less parallel to the primary slip plane (fig. 5.7) and short dislocation walls roughly perpendicular to the slip planes (fig. 5.8) (Steeds, 1966). The density of secondary dislocations in stage I is small compared to that of primary dislocations whereas in stage II the two densities are of the same order. The secondaries tend to occur in the regions containing high densities of primaries; the dislocation mats contain both primary and secondary dislocations, and their interaction products – including Lomer–Cottrell dislocations. The alternate black and white contrast of the mats is caused by rotation in opposite directions at adjacent mats, which in turn is

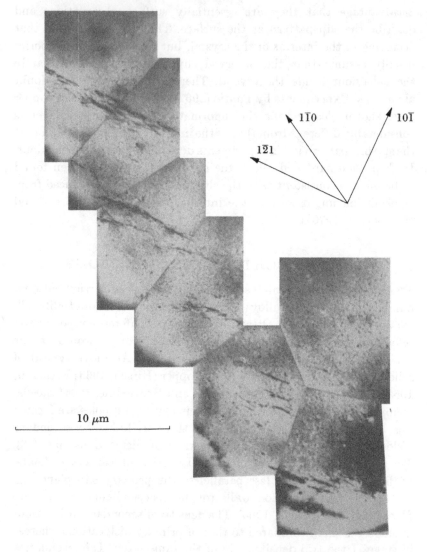

Fig. 5.5. Transmission electron micrograph of section parallel to primary slip plane, showing bands of edge dislocation dipoles in a single crystal of copper deformed to the end of stage I at room temperature. (Steeds, 1966)

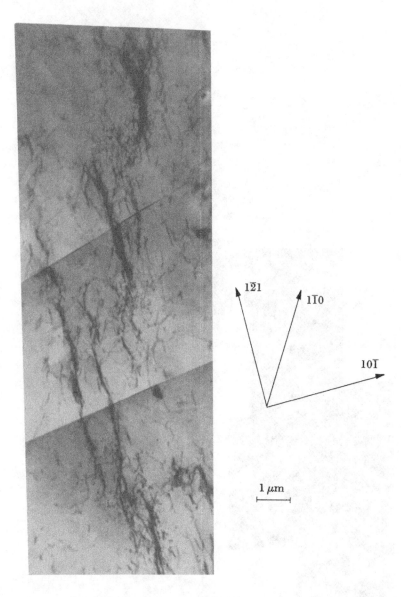

Fig. 5.6. Transmission electron micrograph of section parallel to slip plane showing dipole bands joined by secondary dislocations forming continuous barriers in a single crystal of Cu deformed into stage II at room temperature. (Steeds, 1966)

Fig. 5.7. Transmission electron micrograph of section inclined to (111) slip plane showing sheets ('carpets' or 'mats') of dislocations parallel to the primary slip plane in a single crystal of copper deformed into stage II. Note the alternating black–white contrast indicating rotations (and dislocations) of opposite sign. (Steeds, 1966)

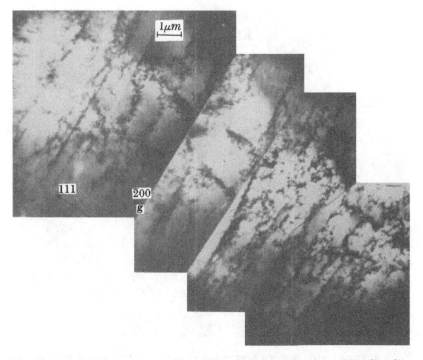

Fig. 5.8. Transmission electron micrograph of section normal to (111) slip plane showing edge multipole walls in a single crystal of copper deformed to the end of stage II. g is the reflection vector. (By courtesy of J. W. Steeds)

caused by dislocations of opposite sign. The mats are rather like imperfect low angle boundaries which are not stress-free (Hirsch and Mitchell, 1968). The structures in niobium are rather similar, both in stage I and stage II (Bowen, Christian and Taylor, 1967; Foxall, Duesbery and Hirsch, 1967); fig. 5.9 shows the mat structure in niobium.

In Cu–Al alloys the dislocations tend to be more exactly confined to the slip planes, this tendency increasing with increasing Al content. In stage I dislocation multipoles are formed, i.e. dislocations of opposite sign on parallel nearby slip planes 'pair-up' with one another; fig. 5.10 shows an example in Cu–10at.%Al. In stage II the primary and secondary slip lines can be recognised quite clearly on the transmission micrographs (fig. 5.11). The secondary slip occurs in bands; in each band, slip on one particular secondary plane predominates. Estimates of dislocation density have shown that in stage I most of the dislocations are primaries and in stage II the density of secon-

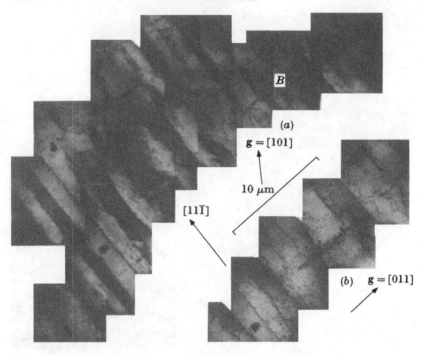

Fig. 5.9. Transmission electron micrograph of section normal to the (011) primary slip plane in a single crystal of niobium deformed into stage II at room temperature, showing sheets of dislocations parallel to the primary slip plane and multipole walls normal to the slip plane; in (b) primary dislocations are out of contrast. **g** is the reflection vector. (Foxall, Duesbery and Hirsch, 1967.) Reproduced by permission of the National Research Council of Canada from *The Canadian Journal of Physics* **45**, 607–29, 1967

daries is $\sim\frac{1}{3}$ of that of the primaries (Steeds and Hazzledine, 1964; Pande and Hazzledine, 1971$a$, $b$).

In stage I for magnesium, typical of a hexagonal metal, the dislocations are mainly in the form of primary edge multipoles (Hirsch and Lally, 1965); forest dislocations threading the primary plane do not seem to be generated. One particular feature common to stage I in Cu, Mg, and Nb is the absence of screws. Observations of different sections of Mg and Cu crystals have shown that the screws have annihilated in the bulk crystal by cross-slip (Steeds and Hazzledine, 1964; Essmann, 1964; Hirsch and Lally, 1965). Presumably, when two screws of opposite sign approach each other on sufficiently close slip planes their interaction can induce cross-slip and annihilation. On the other hand, in the Cu–10at.%Al alloys screw multipoles also exist, and this fact and the tendency for dislocations to be confined

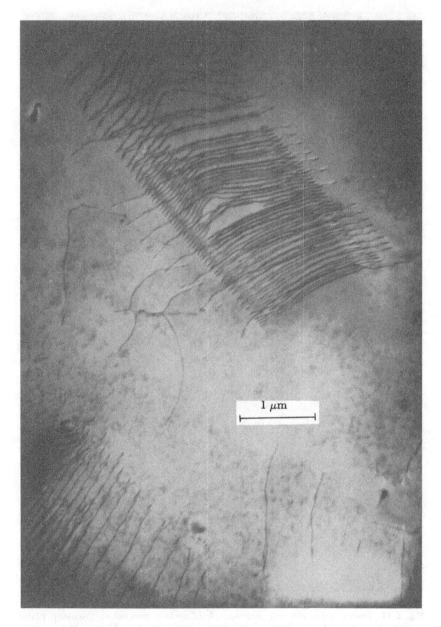

Fig. 5.10. Transmission electron micrograph of section parallel to slip plane showing edge dislocation multipoles in a single crystal of Cu–10at.%Al deformed into stage I at room temperature. (Pande and Hazzledine, 1971a)

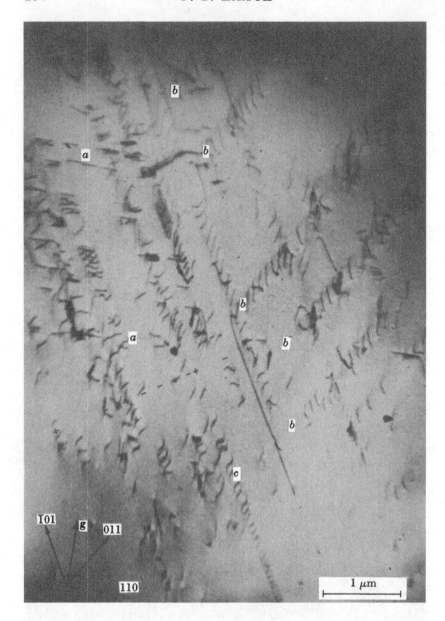

Fig. 5.11. Transmission electron micrograph of section parallel to cross-slip plane showing primary and secondary slip systems operating in stage II in a single crystal of Cu–10at.%Al. g is the reflection vector; a and b indicate slip lines on the critical and conjugate systems respectively; c shows a band of dipoles on the primary system; the signs of the dislocations on the primary system are indicated as + and −. (Pande and Hazzledine, 1971b)

exactly to their slip planes is thought to be due to the lower stacking fault energy in these alloys and consequent difficulty in effecting cross-slip; the increased friction stress due to solution hardening has a similar effect (Taylor and Christian, 1967b).

In stage I of the dispersion-hardened alloys quite different structures are observed. Fig. 5.12 shows a transmission micrograph of a copper crystal containing alumina particles, deformed 10 per cent at 77 °K. Rows of prismatic loops (with the primary Burgers vector) are generated (and left) at the particles by a process of cross-slip which will be described in section 5.6. These loops interact with dislocations with screw components, forming the helices which can be seen in fig. 5.12 (Humphreys and Martin, 1967; Humphreys and Hirsch, 1970). In dispersion-hardened alloys with a Cu–Zn matrix, Orowan loops and prismatic loops are found. Fig. 5.13 shows Orowan loops in Cu–30wt.%Zn with alumina particles, deformed 6 per cent at 77 °K, and fig. 5.14 shows Orowan loops and rows of prismatic loops at larger strains (10 per cent) and 293 °K; the interaction with screws to form helices can be clearly seen.

Seeger (1963, 1968) has criticised the use of the electron microscope to study dislocation arrangements on the grounds that rearrangement can occur during sample preparation. Some rearrangement undoubtedly takes place (Ham, 1962; Valdrè and Hirsch, 1963; for a review see Hirsch, 1963). To reduce the possibility of rearrangement Essman (1963, 1965a, b) has examined specimens neutron-irradiated after deformation but before thinning. It turns out that there is little significant difference between the distributions observed with and without irradiation, except that the irradiated specimens contain in the relatively dislocation-free areas some relatively long dislocations which are bowed out to a radius corresponding to a long-range stress of the order of one half to the full magnitude of the yield stress. Such bowed-out dislocations are also found in dilute Cu–Al alloys in which the friction stress is sufficient to pin the dislocations (Pande, 1968).

The dislocation arrangements observed in this way correspond of course to the unloaded state; Young and Sherrill (1967), using the Borrmann X-ray topograph technique, showed that in copper just below the yield stress the dislocation arrangement was rather different with and without the load applied; in particular, with the stress applied the dislocations are predominantly screws, while in the unloaded state few screws are observed. Similar experiments by

Fig. 5.12. Transmission electron micrograph of section parallel to primary slip plane showing rows (parallel to the primary Burgers vector) of primary prismatic loops at the particles, helices, and dipole clusters in a single crystal of copper containing alumina particles deformed to a shear strain of 15 per cent in stage I at 77 °K. **b** indicates the direction of the primary Burgers vector. (Humphreys and Hirsch, 1970)

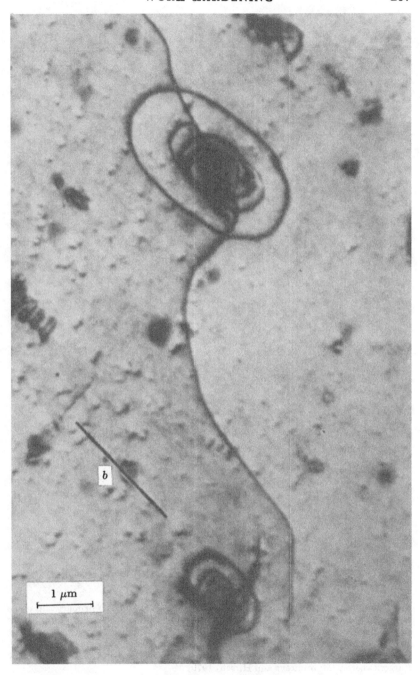

Fig. 5.13. Transmission electron micrograph of section parallel to primary slip plane showing Orowan loops around alumina particles in a single crystal of Cu–30wt.%Zn containing alumina particles, deformed 6 per cent in stage I at 77 °K. **b** indicates the direction of the Burgers vector. (Humphreys and Hirsch, 1970)

Fig. 5.14. Transmission electron micrograph of section parallel to primary slip plane showing primary dislocations interacting with rows of prismatic loops to form helices at particles in a single crystal of Cu–30wt.%Zn with alumina particles, deformed 10 per cent in stage I at room temperature. Note the multipoles at $A$, screw dislocation and a helix at $B$, and an Orowan loop at $C$. **b** indicates the direction of the primary Burgers vector. (Humphreys and Hirsch, 1970)

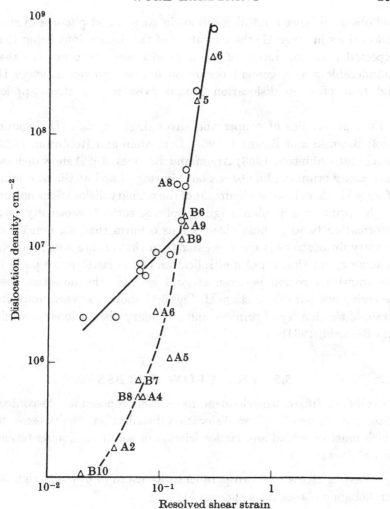

Fig. 5.15. Dislocation density as a function of shear strain in single crystals of copper deformed at 4.2 °K, determined by counting etch pits after subsequent etching treatment at room temperature; ○ primary dislocation, △ forest dislocations. The break in the curve for primary dislocations occurs at the beginning of stage II. (Basinski and Basinski, 1964)

Crump and Young (1968), but using electron microscopy, showed that considerable rearrangement also occurs after the yield, and at relatively low stresses in stage I, but that at stresses of about $\frac{1}{2}$ kg mm$^{-2}$ the rearrangement is negligible. Mughrabi (1968, 1971) carried out such experiments on copper deformed at 78°K in stages I and II and observed bowed-out dislocations in stage I and pile-ups of such dislocations in stage II, the curvature of the dislocations being that expected from the action of the applied stress. He concludes that considerable rearrangement occurs on unloading even in stage II, and that piled-up dislocation groups exist in the stress-applied state.

Etch pit studies of copper and silver single crystals (Livingston, 1962; Basinski and Basinski, 1964; Levinstein and Robinson, 1963; Sabol and Robinson, 1968; Argon and Brydges, 1968) show dislocations along primary glide bands both in stage I and at the beginning of stage II; there is some alignment of the primary dislocations normal to the primary slip planes (glide polygonisation), especially near deformation bands. These observations confirm that the density of primary dislocations is much greater than that of forest dislocations in stage I, but that rapid multiplication of the latter takes place in the transition region between stage I and II, the densities then becoming comparable in stage II. Fig. 5.15 shows the variation with strain of the density of primary and secondary dislocations (Basinski and Basinski, 1964)

## 5.5  THE FLOW STRESS

A number of different mechanisms have been proposed for the control of the flow stress of a given dislocation distribution. The stresses, to which must be added any lattice friction or solute hardening terms, are as follows:

1. Passing stress $\tau$ of two parallel dislocations of opposite sign on parallel glide planes (at a spacing $h$)

$$\tau = \frac{Gb}{4\pi Kh} \tag{5.2}$$

where $G$ is the shear modulus, $b$ the modulus of the Burgers vector, $K = 1$ for screws and $2(1 - \nu)$ for edges, and $\nu$ is Poisson's ratio.

2. Long-range stress from dislocation pile-ups. Assuming the distance $h$ at which the stress is calculated is large compared with the length of the pile up, it can be treated as a super dislocation of Burgers vector with modulus $nb$; thus

$$\tau = \frac{nGb}{4\pi Kh}. \tag{5.3}$$

3. Stress to operate a Frank–Read source (length $l$)

$$\tau = Gb/l. \tag{5.4}$$

4. Stress to overcome attractive or repulsive elastic interaction with forest dislocations spaced at distance $l_f$

$$\tau = \alpha_f \frac{Gb}{l_f} \tag{5.5}$$

where $\alpha_f$ is a constant $\frac{1}{5} \rightarrow \frac{1}{3}$ (Saada, 1960, 1961).

5. Stress to pull out dipoles from sessile jogs spaced at distance $l_j$

$$\tau = \frac{U_D}{b^2 l_j} \tag{5.6}$$

where $U_D$ is the energy per unit length of the dipole; writing $U_D = \alpha_j Gb^3$ we obtain

$$\tau = \alpha_j \frac{Gb}{l_j} \tag{5.7}$$

where for a vacancy producing jog of one atomic plane height, $\alpha_j$ is typically $\sim\frac{1}{5}$.

6. The stress necessary to form a jog at a forest dislocation.

7. The stress necessary for a jog to create a vacancy and to separate from it by conservative glide, or by diffusion of the vacancy.

Mechanisms 1–5 give rise to the so-called temperature independent contribution to the flow stress, $\tau_g$. Mechanisms 6, 7 can be aided by thermal activation, giving rise to the temperature dependent contribution $\tau_s$. It is usually assumed (Seeger, 1957) that the total flow stress $\tau$ is given by

$$\tau = \tau_g + \tau_s. \tag{5.8}$$

The relative magnitude of $\tau_s$ and $\tau_g$ can be estimated from measurements of the change in flow stress when the temperature is changed (Adams and Cottrell, 1955; Cottrell and Stokes, 1955; Basinski, 1959; Seeger, Diehl, Mader and Rebstock, 1957); these suggest that for f.c.c. metals $\tau_g > \tau_s$ in stage II; moreover $d\tau_s/d\tau_g = $ constant in stage II; this latter relation is the so-called Cottrell–Stokes law and has also been studied by a number of workers using change of strain-rate tests (Basinski, 1959; Thornton, Mitchell and Hirsch, 1962). Such a law is not obeyed in b.c.c. crystals, where the temperature dependent term due to lattice friction is large. In stage I in f.c.c. metals the situation is less clear since impurities seem to control $\tau_s$ in this region (Basinski, 1959; Basinski and Basinski, 1964; Gallagher, 1964). However the indications are that in f.c.c. metals the greater the purity the more nearly does $\tau_s/\tau_g$ in stage I approach the value in stage II. The existence of the Cottrell–Stokes law implies either that the obstacles controlling $\tau_g$ and $\tau_s$ are proportional to one another, i.e. that the scale of the dislocation arrangement changes with deformation but not its nature (Cottrell and Stokes, 1955), or that the same obstacle, such as a forest dislocation, is responsible for both the long-range and short-range interaction (Mott, 1955). Jackson and Basinski (1967) have shown in a series of important experiments on copper that the same Cottrell–Stokes ratio is found when slip takes place on the primary system as when the stress axis is changed and slip takes place on a different system; experiments in which crystals are alternately twisted and extended give the same result. Similar results have been obtained by Kocks and Brown (1966) on aluminium. Since the dislocation arrangements in all these cases are rather different from each other the results suggest that the flow stress in stage II, and perhaps even in stage I in pure f.c.c. crystals, is controlled by forest interactions, or that the long-range stress and forest contributions are always in a constant ratio independent of the dislocation distribution.

It should also be noted that mechanisms 1–5 give rise to a flow stress which takes the form

$$\tau \propto \frac{Gb}{l^*} \qquad (5.9)$$

where $l^*$ is some appropriate dimension related to the dislocation arrangement (this is obvious for 1–4; for 5 it is also true if the jog density is related to the forest density). If the dislocation arrangement is in the form of a random network (5.9) can be rewritten

$$\tau = \alpha Gb \sqrt{\rho} \qquad (5.10)$$

where $\rho$ is the *average* dislocation density and $\alpha$ is a constant depending on the interaction. Such a correlation has in fact been found in a number of experiments (Bailey and Hirsch, 1960; Livingston, 1962; Hordon, 1962; Venables, 1962; Basinski and Basinski, 1964; Hirsch and Lally, 1965; Brydges, 1967; Argon and Brydges, 1968), the constant $\alpha$ varying between about 1 and 0.05 depending on the material and the type of dislocation considered. Theoretical estimates of $\alpha$ are not very different for various interactions and for this reason (5.10) cannot be used to distinguish between one mechanism and another. The Stuttgart school (Seeger, Diehl, Mader and Rebstock, 1957; Seeger, Kronmüller, Mader and Träuble, 1961; Seeger, 1968) believe that the flow stress in stages I and II of f.c.c. and in stage I of h.c.p. crystals is controlled by interaction between primaries, i.e.

$$\tau = \alpha G b \sqrt{\rho_p} \qquad (5.11)$$

while other workers have suggested that the forest dislocations are controlling in stage I and stage II of f.c.c. crystals, i.e.

$$\tau = \alpha G b \sqrt{\rho_f} \qquad (5.12)$$

(Basinski, 1959; Hirsch, 1959b; Jackson and Basinski, 1967; Brydges, 1967). Correlations have been attempted between measured flow stress and dislocation densities derived from etch pit observations (Basinski and Basinski, 1964; Jackson and Basinski, 1967; Brydges, 1967; Argon and Brydges, 1968). These suggest that there is a better correlation between flow stress and forest density than between flow stress and primary density, at least in stage I and early stage II of f.c.c. metals. Hirsch and Lally (1965), however, have obtained good correlation between $\tau$ and $\rho_p^{1/2}$ in stage I of magnesium single crystals. It should be noted that such correlations with *average* dislocation densities may not be physically significant tests of the flow stress mechanism for the latter may be controlled by *local* interactions and, since the dislocation distribution is generally not uniform or random, these *local* interactions may not be related in a simple way to average dislocation densities. Nevertheless, this evidence, together with that from Cottrell–Stokes experiments, suggests either that the flow stress in f.c.c. metals is controlled by forest interactions or that the long-range stress and forest contributions are in a constant ratio.

There are, however, cases where other mechanisms may be important. Temperature cycling experiments carried out between room temperature and a higher temperature, after deformation at the higher temperature, have shown a pronounced drop in the flow stress for aluminium at about 200 °C (Hirsch and Warrington, 1961); similar but much smaller decreases in flow stress are found for other metals (Gallagher, 1967). These experiments suggest that for dislocation structures introduced at high temperatures, part of the flow stress depends strongly on temperature. In this case the controlling mechanism may be the dragging force of jogs on screws, leading to the formation of point defects or dipoles, and the jogs may be pinned by impurities. These experiments, however, do not necessarily contribute any information on the interactions important in dislocation structures formed by low temperature deformation.

Another example where other interactions are important is the case of low temperature deformation of dispersion-hardened alloys, which will be discussed in section 5.6.

As stated above, as regards the low temperature flow stress in single crystals of pure f.c.c. metals the evidence seems to favour either the forest mechanism or a constant proportionality between the primary dislocation long-range stress in stage II and forest contributions in stages I and II. The Stuttgart electron microscope observations have demonstrated the existence of pile-ups in the stress-applied state in stage II (Mughrabi, 1968), and have shown evidence for long-range stresses $\sim\frac{1}{2}\tau$ to $\tau$ in the unloaded state (Seeger, 1968). The dependence on deformation of magnetisation at large fields in ferromagnetic metals and alloys has been used to provide further evidence for the existence of long-range stresses in stage II and for their absence in stage I (Kronmüller, 1959; Kronmüller and Seeger, 1961; Seeger and Kronmüller, 1962; Kronmüller, 1967). It should be pointed out that long-range stresses may exist in certain regions on various hardening models in which the dislocation distribution is non-uniform (for discussion see Hirsch and Mitchell, 1967), and that the existence of long-range stresses (inferred many years ago from X-ray line-broadening experiments) does not necessarily constitute proof of the long-range stress model of the *flow stress*; thus, for example, regions of negative long-range stress may be the obstacles at the ends of new slip lines (see below).

Before concluding this section it should be remarked that the definition of flow stress has not been discussed. This problem has been tackled by Kocks (1966) and Foreman and Makin (1966) for the case of dislocations moving through random arrays of point obstacles on a slip plane, corresponding, for example, to forest dislocations intersecting the slip plane. A well-defined flow stress can be derived corresponding to the condition that a dislocation will move indefinitely through the crystal. These treatments yield, for example, the values of $\alpha_f$ to be used in (5.5) for given obstacle strength (see also chapter 4). Kock's theory will be discussed further in section 5.6.

## 5.6  WORK HARDENING THEORIES

The aim of any successful work hardening theory must be to explain not only the flow stress at any given strain when a crystal is deformed in a specific manner but also the nature of the dislocation arrangement and its variation with strain. Most theories, however, attempt only to explain the hardening rates, while making certain assumptions about the dislocated state. Furthermore, in the relation for the strain rate

$$\dot{\epsilon} = \rho_m v b \qquad (5.13)$$

where $\rho_m$ is the mobile dislocation density and $v$ the dislocation velocity, most theories (with one exception – see below) implicitly assume that $v$ is a sensitive function of $\tau - \tau_g$, where $\tau_g$ is the work hardening stress (the 'temperature independent' component), and are concerned with calculations of $\tau_g$, while $\rho_m$ is considered to be effectively constant. We shall limit the discussion to various theories proposed for stages I and II of the single crystal hardening curve; a theory for stage III has been proposed and discussed by Kulmann-Wilsdorf (1968) and by Kocks, Chen, Riguey and Schaefer (1968), but will not be considered here. We shall also not consider work hardening in ordered alloys (e.g. Schoeck, 1969), irradiated metals (Diehl and Hinzner, 1964), or b.c.c. metals at low temperatures when 3-stage hardening is not observed (see Christian, 1970 for review).

### 5.6.1  Stage I

*Metals and solid solutions.* Most of the theories involve the passing stress of two dislocations on parallel glide planes (Seeger, Kronmüller, Mader and Träuble, 1961; Hirsch and Lally, 1965; Hazzledine, 1967; Argon and East, 1968). In Seeger's theory a dislocation emitted by a source is trapped by a dislocation of opposite sign on a parallel plane after it has travelled some distance $L_0$; the back stress at the source is considered to be due to the dislocation emitted by that source, the stresses of all other dislocations being neglected. The condition for flow is that the stress is increased during the strain interval corresponding to the slip over a distance $L_0$ by an amount exceeding the back stress so that the source can emit another dislocation; this latter can help the first dislocation to bypass the dislocation of opposite sign at which it is trapped. The next dislocation is then trapped and the process is repeated. Assuming that the slip line length $L$ and the slip line spacing $h$ are constant, the work hardening rate $\theta_\mathrm{I}$ is found to be

$$\theta_\mathrm{I} = \frac{8G}{9\pi} \left(\frac{h}{L}\right)^{3/4}. \tag{5.14}$$

The observed values of $\theta_\mathrm{I}$, $h$, $L$ for a number of metals and alloys are found to be consistent with this relation. However, it seems that further justification is needed for the assumption made in the theory that the effective back stress at the source is that due to the dislocation emitted by the source, when in fact it is trapped by a dislocation of opposite sign to form a dipole. Certain other difficulties in connection with this theory have been discussed by Nabarro, Basinski and Holt (1964). It should be noted that the theory is designed for a crystal in which $h$ and $L$ are constant and the number of dislocations per slip line increases with increasing strain, as is approximately the case for Cu, Ni–Co and Zn crystals, for example.

Because of the obvious pairing of dislocations into dipoles or multipoles, theories in terms of such arrangements have been developed. In the theory of Hirsch and Lally (1965) (applied to magnesium) the flow stress is simply related to the observed dislocation arrangement in the form of edge multipoles and is thought to be effectively the stress needed by a dislocation to pass a small number of excess dislocations within each multipole. The flow stress is then related to the observed dislocation density and reasonable agreement with the observed parabolic stress–strain curve is obtained. As in Seeger's theory, this theory also only predicts a correlation and does

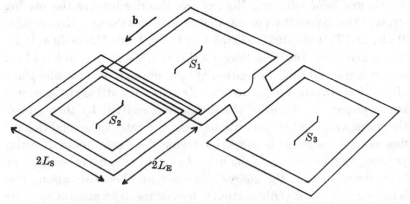

Fig. 5.16. Dislocation sources interacting producing edge and screw dipole bands; the screws annihilate by cross-slip. (Hirsch and Lally, 1965)

not explain, for example, the slip line spacing at a given strain, the number of dislocations per slip line etc.

Following Tetelman (1962), the mechanism proposed for the formation of multipoles is shown in fig. 5.16 (Hirsch and Lally, 1965). Dislocations from sources $S_1$, $S_2$, $S_3$, meet as shown, and if the slip plane spacing is small enough the screws annihilate by cross-slip and the edges form multipoles. The density of edge dislocations, $\rho_E$, produced in this way is related to the slip line length $L_E$ by the relation

$$\frac{d\rho_E}{d\epsilon} = \frac{1}{L_E b}. \tag{5.15}$$

In magnesium the densities of dislocations observed fit in well with a constant slip line length $L_E$ of reasonable magnitude (see Hirsch and Lally, 1965).

The difficulty with this theory is that it is not clear why the multipoles should cause any hardening by themselves since the applied stress acting on one extra dislocation joining the group will just cause the whole group to move along (Washburn, 1963; Sharp, Makin and Christian, 1965). Furthermore, excess dislocations of one sign on one of the slip planes can, by a process of interchange, effectively pass the multipole easily (Hazzledine, 1967). Any hardening must be due to the difficulty of moving such multipoles, possibly due, for example, to the drag of the jogs at their ends (Sharp, Makin and Christian, 1965), interaction with point defects, Peierls stress, or forest dislocations; in the case of the magnesium crystals referred to above, the resistance to glide may have been provided by particles which were known to be present. Effectively, therefore, *some other hardening mechanism* must be present to make the hardening from multipoles effective.

Argon and East (1968) have developed a statistical theory of hardening in stage I which is also based on the formation of multipoles; these are thought to be nucleated at grown-in forest dislocations, to grow with increasing strain by trapping more edge dislocations and thereby to cause hardening. The theory, which predicts that $\tau \sim \epsilon^{1/4} \to \epsilon^{1/3}$, assumes that the process controlling the dislocation velocity $v$ is the intersection of forest dislocations, that stresses from multipoles have no effect on $v$, and that the average $v$ is constant; i.e. in equation (5.13) the velocity $v$ is controlled by forest intersections and, assuming that the relation (5.12) holds (which is experimentally observed for copper by Argon and Brydges, 1968), $v$ is constant, independent of $\tau$ and $\epsilon$. Now the mobile dislocation density $\rho_m$ is controlled by the multipole trapping mechanism, and for a given $\tau$, $\rho_m$ decreases with increasing $\epsilon$. Thus in order to maintain a constant $\dot{\epsilon}$ with increasing $\epsilon$ for a constant $v$ the stress $\tau$ must increase to maintain $\rho_m$ at the appropriate value. It should be emphasised that the theory of Argon and East is rather different from those of Seeger *et al.* (1961), Hirsch and Lally (1965), and Hazzledine (1967) in that in the former the hardening by multipoles controls $\rho_m$ whereas in the latter theories the hardening effectively controls the stress dependent velocity, it being assumed that work hardening has a more important effect on $v$ than on $\rho_m$. Thus the stress dependence of the average velocity of dislocations in f.c.c. metals (assuming it to be controlled by forest intersections) is such that if the forest density remained constant with increasing strain, any increase in stress necessitated by the effect of multipole trapping on $\rho_m$ would lead to a very large increase in $v$, so that at constant $\dot{\epsilon}$ the hardening rate would be extremely small. It is therefore important to realise that in the theory of Argon and East relation (5.12) is *assumed* to hold, in agreement with experiment, so that the forest density increases at the correct rate to keep $v$ constant; no explanation is offered to justify this relationship. By contrast, in the other theories of hardening due to primary dislocations and multipoles $v$ is assumed to be a function of $\tau - \tau_g$, where $\tau_g$ is the work hardening; this $\tau_g$ in the velocity term is then calculated directly as a function of $\epsilon$, as the temperature independent contribution to $\tau$.

In Hazzledine's (1967) theory the number of sources is assumed constant, dislocations from sources on different slip planes travel until they trap each other to form multipoles, and at a later stage in the stress–strain curve, i.e. at a higher stress, the dislocations can bypass each other again and move on until they are trapped by another group of dislocations. Each time the group is released the source emits more dislocations to fill up the slip line with multipoles. The theory is designed to apply to alloys with high friction stresses which will stabilise the multipoles; it has been worked out in detail for the two-dimensional case in which screw dislocations annihilate by cross-slip instead of forming multipoles, leaving only edge multipoles (see fig. 5.16). The essential features of the theory can be derived in a simplified way as follows: suppose that the density of source dislocations is $\rho_0$ and that the mean distance travelled by the edge dislocations emanating from a source is $\bar{L}$; then if each source gives out an average of $2n$ dislocations at a strain $\epsilon$,

$$\epsilon = 2n\rho_0 \bar{L}b. \qquad (5.16)$$

Hazzledine (1966) has shown that, assuming that each dislocation from a source $S_1$ is paired with a dislocation of opposite sign from $S_2$ (see fig. 5.17) to fill the slip line with multipoles, the dipoles within the multipoles will be equally spaced and for edge dislocations the slip line length

$$L = 2\bar{L} = 6nh \qquad (5.17)$$

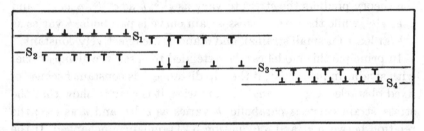

Fig. 5.17. Dislocations from different sources forming multipoles. (After Hazzledine, 1967.) Reproduced by permission of the National Research Council of Canada from *The Canadian Journal of Physics* 45, 765–75, 1967

where $h$ is the mean slip line spacing. If the crystal is filled uniformly with multipoles

$$4\bar{L}h\rho_0 = 1 \qquad (5.18)$$

and the stress $\tau_g$ will be

$$\tau_g = \tau_0 + \frac{Gb}{8\pi(1-\nu)(2h)} \qquad (5.19)$$

where $\tau_0$ is the friction stress in the undeformed crystal, e.g. due to solute. Equation 5.19 gives the stress necessary for dislocations to bypass one another in a multipole of spacing $2h$ ($h$ is the average slip line spacing between 0 and $2h$ within which range dislocations will trap each other at stress $\tau$). It follows that

$$\tau_g - \tau_0 = \frac{G}{8\pi(1-\nu)} (3\rho_0 b^2)^{1/3} \epsilon^{1/3} \qquad (5.20)$$

which agrees to within a small factor with Hazzledine's more detailed treatment. If $\rho_0$ remains constant this predicts a stress–strain curve of the form $\tau_g - \tau_0 \propto \epsilon^{1/3}$. (A treatment based on a somewhat similar model to Hazzledine's predicts $\tau_g - \tau_0 \propto \epsilon^{1/7}$ (Yazu, 1968).) The edge dislocation density $2n\rho_0$ is related to $\tau_g$ by the equation

$$\tau_g - \tau_0 = \alpha G b \sqrt{\rho} \qquad (5.21)$$

where

$$\alpha = (3/2)^{1/2}/8\pi(1 - \nu) \sim 0.073. \qquad (5.22)$$

Hazzledine shows that (5.21) explains the observed dependence of $\rho$ on $\tau_g$ for magnesium (Hirsch and Lally, 1965), and the unusually small value of $\alpha$ (the experimentally observed $\alpha \sim 0.05$). However, the theory predicts the stress to vary as $\epsilon^{1/3}$, $h$ as $\epsilon^{-1/3}$, $n$ as $\epsilon^{2/3}$ and $\bar{L}$ as $\epsilon^{1/3}$, while the actual stress–strain curve is parabolic, $h$ varies as $\epsilon^{-1}$ (at least for small strains), and $n$ and $\bar{L}$ are effectively constant.

In principle this model can be extended to cases in which $\rho_0$ varies with strain. In particular, if the slip distance $L$ is constant because of fixed obstacles, e.g. low angle boundaries, it is easy to show that the stress–strain curve is parabolic, $h$ varies as $\epsilon^{-1/2}$, and $n$ as $\epsilon^{1/2}$; the relation between $\tau_g$ and $\rho$ (equation 5.21) remains unchanged. If the slip planes also contain obstacles which are penetrable by dislocations but which can trap dipoles – for example forest dislocations – the multipoles can be trapped even if the friction stress is low, so the multipole theory should also be applicable to that case.

In the case of magnesium another mechanism may be operating. Electron micrographs sometimes show rows of prismatic loops behind particles in the slip plane (Hirsch and Lally, 1965). The 'pure crystal' behaves, therefore, like a very weak dispersion-hardened alloy. As will be shown below for dispersion-hardened alloys, the stress–strain curve is parabolic, as observed, and for the theory presented, $h \propto \epsilon^{-1/2}$ and $n \propto \epsilon^{1/2}$, as for the multipole theory. Unfortunately the volume fraction of particles is not known so that quantitative comparison with theory is not possible. It appears, then, that for magnesium the stress–strain curves can be explained either by the multipole mechanism or by the dispersion hardening mechanism; however, the observed dependence of $h$ and $n$ on strain remain to be explained; it should be noted that $h$ and $n$ are determined from surface observations and therefore open to doubt (see section 5.3).

It is clear that Hazzledine's multipole model makes a number of simplifying assumptions (for a discussion see Hazzledine, 1967). For example it is assumed that equal numbers of dislocations are emitted by the pair of sources whose dislocations are forming a multipole. In general the numbers of dislocations will be unequal and the excess number will effectively glide away from the multipole (by a process of interchange), presumably forming multipoles with dislocations from other sources. This means that dislocations on parallel slip planes can pass one another even though the passing stress for equal numbers of dislocations is not exceeded. Furthermore the dislocations emitted on opposite sides of the source may be trapped at different distances from the source; the number emitted will be the smaller of the two numbers of dislocations needed to fill the slip line on the two sides; thus the slip lines will not be filled with multipoles. Also, if a multipole on one side of the source decomposes, the source may be unable to emit further dislocations because of the multipole still trapped on the other side; this again results in the slip line not being filled with multipoles. This is indeed in agreement with electron microscope observations of dislocation distributions (Hazzledine, 1967). All these mechanisms affect (5.17), suggesting that this equation is not generally applicable.

We shall now return to the discussion of the work hardening mechanism in the case when the slip line length and spacing remain constant and the number of dislocations per slip line increases with strain; Seeger's theory (Seeger, Kronmüller, Mader and Träuble, 1961, see above) is designed to explain this case. We have seen that dislocations of opposite sign on parallel slip planes can effectively pass one another at the critical passing distance at stresses much lower than the passing stress (Hazzledine, 1967); an excess dislocation of one sign arriving at a multipole will push away a dislocation of the same sign on the other side, or, if the friction stress is low, the whole or part of the multipole can be swept along. Alternatively, the multipole may be destroyed (Neumann, 1971; Hazzledine, 1971b); which of these interactions occurs depends on the position of the slip plane of an incoming dislocation relative to those of the existing multipole. Thus the active slip line spacing can be smaller than that expected from the passing stress. This explains why the observed slip line spacing in copper and other f.c.c. metals in stage I (Mader, 1957, 1963) is several times smaller than the critical passing distance, and why the slip lines are not filled with multipoles. We expect however groups of dislocations to be trapped in the form of multipoles at obstacles such as forest dislocations (see fig. 5.18). Assuming that each multipole consists of $n$ dislocations of either sign and that the spacing

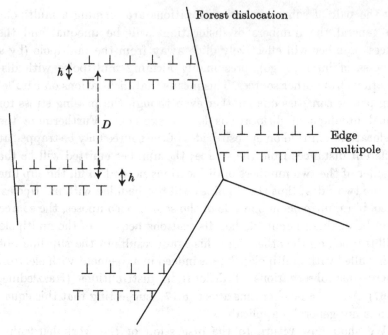

Fig. 5.18. Multipoles trapped at forest dislocations

between the two dislocations in any dipole within the multipole is $h$, the maximum internal stress $\tau_g$ on a slip plane between the (edge) multipoles (spaced at distance $D$) is

$$\tau_g \sim \frac{nGb}{2\pi(1-\nu)}\left(\frac{1}{D} - \frac{1}{D+h}\right)$$

$$\sim \frac{nGbh}{2\pi(1-\nu)D^2}. \qquad (5.23)$$

The dislocation density $\rho \sim 2n/D^2$, and using (5.15) with $L_E$ constant we find

$$\tau_g = \frac{Gh\epsilon}{4\pi(1-\nu)L_E}. \qquad (5.24)$$

Now if the total stress applied is $\tau$, only multipoles with values of $h$ satisfying the passing stress (5.19) will be stable; therefore if $\tau_f$ is the forest stress and $\tau_0$ the friction stress (e.g. due to solute hardening),

$$\tau = \tau_0 + \tau_f + \tau_g = \tau_0 + \tau_f + \frac{G^2\epsilon b}{64\pi^2(1-\nu)^2 L_E(\tau-\tau_0)}. \qquad (5.25)$$

Hence,

$$\tau - \tau_0 = \tfrac{1}{2}\tau_f + \tfrac{1}{2}\tau_f \left( 1 + \frac{G^2 b\epsilon}{16\pi^2(1-\nu)^2 L_E \tau_f^2} \right)^{1/2}. \qquad (5.26)$$

For small hardening rates

$$\tau - \tau_0 = \tau_f + \frac{G^2 b\epsilon}{64\pi^2(1-\nu)^2 L_E \tau_f}. \qquad (5.27)$$

This gives linear hardening if $\tau_f$ remains constant. On the other hand, if $\tau_f$ increases e.g. linearly, with strain, the work hardening rate will decrease slowly with strain to a constant value determined by the forest hardening term only. Putting $\tau_f$ equal to the yield stress in a pure metal ($\sim G/10^5$) and $L_E \sim 0.5$ mm, $\theta_I/G \sim 10^{-4}$, which is of the correct order of magnitude. The strong orientation dependence of $\theta_I$ (see Mitchell, 1964) could be due to an orientation dependence of the rate of increase of $\tau_f$ with strain, or of the slip line length $L_E$.

This model leaves open the possibility that at sufficiently large strains in stage I the flow stress is controlled mainly by forest dislocations, in agreement with etch pit observations on copper (Basinski and Basinski, 1964; Argon and Brydges, 1968).

With regard to the multiplication of forest dislocations in stage I (as found in copper), no convincing explanation has yet been given for the observed rate. However, some multiplication is likely to occur simply because the applied stress increases due to work hardening of the primary system. Furthermore, the internal stresses near multipoles, although small, may also generate some forest dislocations; the more effective source of internal stresses may, however, be that due to 'loose' primary dislocations not trapped in multipoles and able to move easily through the crystal. Whichever mechanism is operating, the rate of multiplication of forest dislocations is expected to be smaller in the presence of a strong friction stress, e.g. due to solute effects, and this correlates with the fact that in solution-hardened alloys $\theta_I$ tends to be less than in pure crystals (see Mitchell, 1964a).

Before summarising this section it should be mentioned that Friedel and Saada (1968) have proposed, for the initial part of stage I, a hardening mechanism in which the dislocations moving through a forest collect jogs and pull out dipoles whose height increases with increasing slip distance. The flow stress then increases logarithmically with strain. This theory cannot apply to most of stage I in those cases for which the slip distance is known to be constant. Furthermore, on reverse straining this theory would predict a very large Bauschinger effect for which there seems to be no evidence.

In conclusion we should like to make the following points:

1. For pure crystals without internal or surface obstacles and with low friction stress, no convincing theory has yet been developed which explains the hardening in terms of interaction between primary dislocations. Dipoles and multipoles in such crystals are so unstable and can be moved around so easily by excess dislocations of one sign that it seems difficult to envisage a convincing hardening mechanism. For primary interaction to be effective is seems essential to have other types of obstacles present or generated (or a large friction stress) which can trap the multipoles.

2. If the multipoles are trapped their small internal stress fields can cause hardening, which, to a first approximation, is linear with strain.

3. If forest multiplication is easy, then the hardening rate may be primarily determined by forest hardening, particularly at larger strains in stage I. There is no theory which predicts the rate of forest multiplication in stage I.

4. If the active slip lines get filled with multipoles, hardening is no longer linear, and the slip line spacing is expected to decrease with increasing strain.

This brief account (in which we have not considered effects due to generation of point defects, formation of prismatic loops, short-range or long-range order, stress inhomogeneities due to grips etc.) suffices to show that hardening in stage I of pure metals and solid solutions is still not well understood.

The principal difficulty lies in the fact that there is doubt about the nature of the important obstacles created during strain and of how they multiply with strain. Recently, however, work hardening theories have been developed for certain types of dispersion-hardened alloys and in this case the obstacles are reasonably well-characterised and the variation with strain can be determined. It should therefore be easier to work out and test the work hardening theory for such alloys.

*Dispersion-hardened alloys.* The stress–strain curves for these alloys have been determined by Ashby (1966), Ebeling and Ashby (1966),

Humphreys and Martin (1967) and Jones and Kelly (1968) (see section 5.2). If the particles are small the dislocation reactions are particularly simple, and the work hardening theory to be described (Hirsch and Humphreys, 1969, 1970b) is designed to apply to this case. Electron micrographs (fig. 5.12) have shown that prismatic loops are generated as a result of interaction of the moving dislocation with the particle; it is likely that a small number of Orowan loops are formed first and that when this number exceeds a certain critical value the innermost Orowan loop collapses to form two prismatic loops by a double cross-slip mechanism, one of the prismatic loops being carried off, in the form of a double jog, by a glide dislocation. The mechanism is illustrated for the case of one Orowan loop in fig. 5.19 (Hirsch,

Fig. 5.19. Generation of a pair of prismatic loops by double cross-slip of an Orowan loop at a particle under the influence of stress of a glide dislocation. One of the prismatic loops interacts with the dislocation to form a double jog and is carried away from the particle

1957; Ashby and Smith, 1960; Humphreys and Martin, 1967; Hirsch and Humphreys, 1969; Gleiter, 1968; Duesbery and Hirsch, 1970; Humphreys and Hirsch, 1970). In the case of copper matrix alloys Orowan loops are not observed, but these are likely to anneal out at room temperature by climb around the particle (Hirsch and Humphreys, 1969); they have been observed in Cu–Zn matrix alloys (figs. 5.13, 5.14). After an initial small strain when only Orowan loops are formed, each glide dislocation leaves one prismatic loop, and in this way a row of loops is formed by a set of dislocations gliding on the same slip plane. The hardening effect of the rows of prismatic loops can be estimated as follows. The glide dislocations coming later in the sequence interact with the row of loops to form a helix (if the dislocation has screw character); this helix interaction must be reversed before the dislocation can move past the row of loops and generate one more loop at the particle. Fig. 5.20 shows how this can be achieved: the screw bows out into an arc, then becomes unstable and 'blows out', and at some critical value of the angle between the side arm and the helix the latter is 'unzipped' leaving the row of loops behind. When the dislocation reaches the particle it forms another loop which adds to the row. As the screw bows out the turns of the helix become compressed like a spring, but when the pitch is $\sim R$ (the radius of the loop) the repulsion between the turns is so strong that little further compression takes place (Grilhé, 1967).

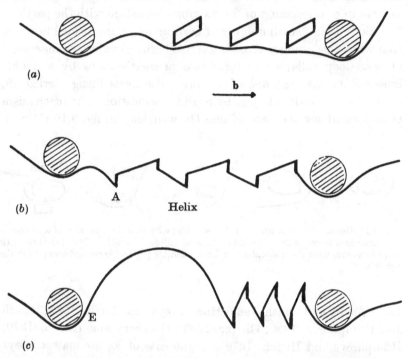

Fig. 5.20. Screw dislocation interacting with a row of prismatic loops forming a helix; to move past the particle the screw must bow out to a critical configuration to 'unzip' the helix, leaving the original row of loops behind and forming another Orowan loop or prismatic loop at the particle. (Hirsch and Humphreys, 1969.) Reprinted from *Physics of Strength and Plasticity* edited by A. S. Argon, by permission of The M.I.T. Press, Cambridge, Massachusetts. © The M.I.T. Press, 1969

Effectively, therefore, the rows of $n$ loops act as parallel linear obstacles of length $nR$. Edge dislocations will not form helices, but will compress the loops again to an interloop spacing $\sim R$ (Bullough and Newman, 1960). With increasing strain $n$ increases and therefore the lengths of the obstacles in the slip plane increase, leading to a 'self hardening' of the slip line. This self hardening law has been determined using the computer program of Foreman and Makin (1966) on the assumption that the critical 'unzipping' angle, i.e. the angle between the side arm and helix in fig. 5.20, is $\pi/2$, except at the particle, where the breaking angle is assumed to be zero (Foreman, Hirsch and Humphreys, 1970). Fig. 5.21 shows the 'self hardening' stress as a function of obstacle length. Above a certain critical length $s_0$ the self hardening flow stress, $\tau_{sh}$, is given by

$$\tau_{sh} = \tau_0' + \frac{1.64Ts}{bD_s^2} = \tau_0' + \frac{1.64Tn_pR}{bD_s^2} \tag{5.28}$$

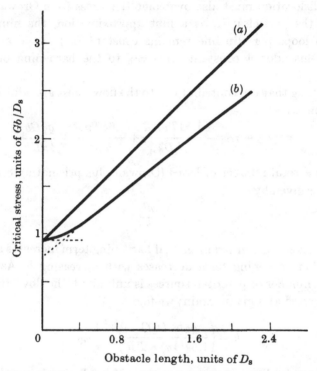

Fig. 5.21. Flow stress as a function of obstacle length in a slip plane, obtained by computer simulation, (a) for infinitely hard obstacles (zero breaking angle), (b) for obstacles with 90° breaking angle except for hard point at one end representing particle. ($G$ is shear modulus, $b$ is the modulus of the Burgers vector, $D_s$ the mean interparticle spacing in the slip plane.) (Hirsch and Humphreys, 1970b)

where $T$ is the line tension, $s$ the obstacle length, $b$ modulus of the Burgers vector, $n_p$ the number of prismatic loops, $D_s$ the mean planar spacing of the particles and $\tau_0'$ a stress slightly smaller than the Orowan stress.

In addition to this self hardening effect, the dislocations must also be able to pass other dislocations on parallel active glide planes; this gives rise to the 'interaction hardening'. The passing stress is taken as

$$\tau_{1h} = \frac{q_2 Gb}{4\pi Kh} \tag{5.29}$$

where $K$ is 1 and $2(1 - \nu)$ for screws and edges respectively, $h$ is the slip line spacing, and $q_2$ is a factor which takes into account that when reaching the critical unzipping configuration the dislocation may not have to overcome the maximum passing stress.

The dislocation must also overcome the stress from Orowan loops around the particles; if, as a first approximation, the number of Orowan loops per slip line remains constant $(n_0)$ above a critical strain, this adds a constant term $\tau_{Oh}$ to the hardening of a slip line.

Assuming that the contributions to the flow stress are additive, the total flow stress is

$$\tau = \tau_0' + \tau_{Oh} - \frac{1.64 T n_0 R}{b D_s^2} + \frac{1.64 T n R}{b D_s^2} + \frac{q_2 Gb}{4\pi Kh} \tag{5.30}$$

where $n = $ total number of loops (Orowan plus prismatic). Since the strain $\epsilon$ is given by

$$\epsilon = \frac{nb}{h} \tag{5.31}$$

we see that at a given strain the self hardening term increases and the interaction hardening term decreases with increasing $h$. Assuming that the number of potential sources is unlimited, the flow stress can be minimised at a given strain; we find

$$h = \left( \frac{q_2 b^3 D_s^2 G}{1.64(4\pi) KTR} \right)^{1/2} \frac{1}{\epsilon^{1/2}} \tag{5.32}$$

and the strain dependent term of the self hardening is equal to the interaction hardening. Substituting in (5.30), this equation can be written in the form

$$\tau - \tau_0 = \left( \frac{1.64 q_2 GTR}{\pi Kb D_s^2} \right)^{1/2} (\epsilon^{1/2} - \epsilon_0^{1/2}) \tag{5.33}$$

where $\epsilon_0$ is the critical strain above which this theory applies, and which takes into account the various constant terms in (5.30). Substituting for the line tension

$$T = \frac{Gb^2}{4\pi K'} \ln\left( \frac{d_s}{r_0} \right) \tag{5.34}$$

where $K' = 1$ for an edge and $K' = (1 - \nu)$ for a screw bowing out and $d_s$ is the mean diameter of the section of the particle intersecting a slip plane, and putting $D_s^{-2} = 3f/2\pi R^2$ where $f$ is the volume fraction, we find that

$$\tau - \tau_0 = G\left(\frac{4.92q_2 \ln (d_s/r_0)}{4\pi^3 KK'}\right)^{1/2} \left(\frac{fb}{2R}\right)^{1/2} (\epsilon^{1/2} - \epsilon_0^{1/2}). \qquad (5.35)$$

The stress–strain curve is therefore parabolic and the hardening rate is proportional to $(fb/2R)^{1/2}$ as was first demonstrated from the experiments by Ashby (1966). For copper with alumina particles at 77°K, Hirsch and Humphreys (1970b) have found that $(d\tau/d\epsilon^{1/2})$ $(fb/2R)^{-1/2}(\ln d_s/r_0)^{-1/2} = 1.0 \times 10^3$ kg mm$^2$, compared with theoretical values for screws and edges of $1.1 \times 10^3$ and $0.8 \times 10^3$ kg mm$^{-2}$ respectively; the agreement is satisfactory. We note from (5.32) that $h \propto \epsilon^{-1/2}$ and that therefore the fraction of particles with loops $p \propto \epsilon^{1/2}$. Transmission microscope observations show $p$ to be of the correct order of magnitude and to increase with strain, the agreement not being perfect but reasonable bearing in mind that the loops may be dispersed along their glide cylinder by interaction with dislocations moving on nearby glide planes. The average number of loops observed per particle, $\bar{n}$, is of the same order as that expected from theory, although a little smaller; it is uncertain whether the small discrepancy ($\bar{n}_{\text{expt}} \sim 25$, $\bar{n}_{\text{theory}} \sim 30$ at 15 per cent strain) is within experimental error or whether it reflects a small number of Orowan loops which have annealed by climb before observation.

It is worth considering the following points about this theory. Firstly, the theory derives from first principles the form of the dislocation arrangement (at least in part), the stress–strain curve and the slip distribution. Secondly, the flow stress is determined by *a local interaction* and not by some *average dislocation density*. Nevertheless, we can put the flow stress into the usual form; the average number of loops per particle is given by $\bar{n} = \epsilon 2R/b$ and since each loop contributes $\sim 2\pi R$ of dislocation line (5.35) can be written

$$\tau - \tau_0 = \alpha Gb(\rho^{1/2} - \rho_0^{1/2}) \qquad (5.36)$$

where $\rho$ is the density of dislocations in the form of loops and

$$\alpha = \left(\frac{1.64q_2 \ln (d_s/r_0)}{8\pi^3 KK'}\right)^{1/2}.$$

For typical values of the parameters, $\alpha \sim 0.25$. Thus we see that the usual flow stress equation is obtained, but it does not say anything about the physical mechanisms controlling the flow stress; equation (5.36) simply gives the dislocation density in the form of loops stored at a given stress. Ashby (1966) derived, in effect, the parabolic stress–strain law of equation (5.35) by using (5.36) (as a flow stress equation) as his starting point and by assuming that the loops are uniformly distributed and that they give rise to a kind of forest hardening. Such a model does not agree with the experimental observations of the microstructure for alloys with small particles, nor can it make predictions about slip line distributions. Furthermore, the constant $\alpha$ is somewhat uncertain. Nevertheless the parabolic form and the dependence on $R$ and $f$ can be derived, as was shown by Ashby, and this indicates at once the usefulness and the limitations of so-called flow stress equations relating flow stress to *average* dislocation density, i.e. describing the dislocation distribution in terms of one parameter only (cf. Kocks, 1966).

It should also be noted that the possibility of making detailed predictions about the stress–strain curve and the slip distribution in this case is due to the fact that a model is available for the 'self hardening' law of a slip line. It is the lack of understanding of the self hardening of a slip line that constitutes one of the main difficulties in the development of work hardening theories for pure metals and solid solutions.

Since this article was prepared a number of relevant papers have been published, including Brown and Stobbs (1971a, b), Hart (1972), Brown (1973), Gould *et al.* (1973), Atkinson *et al.* (1973), and a recent review of the deformation of dispersion-hardened alloys is given by Ashby (1971). In particular, the large Bauschinger effect in dispersion-hardened alloys (Wilson, 1965) can be explained in terms of the stresses from Orowan loops remaining around the particles (Brown, 1973).

### 5.6.2   Stage II

In considering the various models proposed it is useful to bear in mind certain basic facts:

1. One of the characteristic features of stage II hardening of pure f.c.c. metals is that new slip lines are formed as the stress increases, that the number of dislocations per slip line remains approximately

constant, and that the slip line lengths decrease with increasing strain. The slip lines are fully formed in a small stress interval. The fact that new slip lines of limited length are formed implies that the crystal contains 'soft' and 'hard' regions: dislocations are generated in some regions of the crystal and are blocked at others. The observed dislocation structure is in fact very inhomogeneous (see section 5.4); presumably the regions of high dislocation density are 'hard', the regions in between 'soft'.

2. If the dislocations on a slip line are stopped at previously formed obstacles, i.e. locally hard regions, then the observed decrease in slip line length with strain implies an increase in the number or strength of these obstacles.

3. The dislocation arrangement chosen must be compatible with the stress at which it is supposed to be formed, and it must be stable when the stress is removed.

4. Since linear hardening is associated with activity on secondary systems, the theory must explain why this slip takes place at all, how it affects the hardening rate, and what contribution it makes to the plastic strain.

Further discussion of these general points is given in Hirsch and Mitchell (1967, 1968).

We shall now discuss the various theories in turn.

*Forest, meshlength, statistical and jog theories.* The basic idea underlying the forest theory (Basinski, 1959; Hirsch, 1959b; Saada, 1960, 1961) is that long-range stresses due to piled-up groups cause slip on secondary systems, producing networks which are stable on removal of the stress and which cause hardening. The flow stress is controlled by the elastic stresses which have to be overcome in intersecting forest dislocations; it is given by (5.12). The geometry of the dislocation arrangement is assumed to remain similar during deformation so that forest ($\rho_f$) and primary ($\rho_p$) densities are proportional to one another, and the slip line length ($L$) is proportional to the mean forest separation, i.e.

$$\rho_f = k_1 \rho_p \tag{5.37}$$

and

$$L = k_2 / (\rho_f)^{1/2}. \tag{5.38}$$

Using (5.12), (5.15), (5.37) and (5.38) we find

$$\theta_{II}/G = \frac{k_1}{4k_2}. \tag{5.39}$$

Now $k_1$ is expected to be $\sim1$ since in the course of stress-relaxation
networks are formed in which the spacing of forest dislocations is of
the same order as that for primaries. Hence, in order to obtain agreement with experiment, $k_2 \sim75$. The attractive feature of the forest
theory is that it predicts the observed Cottrell–Stokes law (Basinski,
1959; Hirsch, 1959b). However there is no satisfactory explanation for
the magnitude of $k_2$. Within the framework of this model, the reason
for this failure is that the dislocation structure is described in terms
of only one parameter, i.e. an average forest separation. There is
then no basic mechanism for blocking a slip line. Any successful forest
theory must be based on a model in which the dislocation distribution
is inhomogeneous, as observed in practice. It should be noted,
however, that Kocks (1966) has shown that when a dislocation moves
through a random array of point obstacles debris is left behind and,
although the glide dislocations in the slip line are not stopped, work
hardening occurs. This theory will be discussed below.

Kuhlmann-Wilsdorf (1962) proposed a theory which is formally
rather similar to the forest theory, the main difference being that the
flow stress in her model is the stress to operate a Frank–Read source.
This theory suffers from the same disadvantage as the forest theory in
that $k_2$ cannot be derived in a convincing manner. Once the flow
stress is exceeded and dislocations are generated by a source it is
difficult to see how slip lines can be blocked if the network is random
(see Hirsch and Mitchell, 1967).

It is appropriate to consider at this stage the statistical theory of
flow stress and work hardening due to Kocks (1966). In this theory a
random distribution of point obstacles is assumed and the propagation
of dislocations through such arrays is considered. It turns out that
(i) there is a critical stress at which a dislocation can move through
a crystal indefinitely – this is the macroscopic flow stress – and (ii)
the dislocation leaves behind concave dislocation loops encircling
small numbers of point obstacles. The first result is directly applicable
to the Orowan stress, and the relation obtained between flow stress
and obstacle spacing has been confirmed by the computer experiments of Foreman and Makin (1966). The second result shows that as
dislocations travel through the crystal they cause work hardening by
generating debris. Although the dislocations cannot be stopped in
the crystal, a mean free path can be defined effectively by equation
(5.15) relating dislocation density and strain (Nabarro, Basinski and
Holt, 1964; Kocks, 1966). Assuming a flow stress relation of the type

of (5.11), a linear work hardening rate is obtained which depends on a parameter rather similar to $k_2$ in the forest theory. This parameter can be calculated and gives the observed work hardening rate in stage II. Kocks makes the important point that strong hardening is produced even though the slip lines are not stopped. He suggests that the observed slip line ends may correspond to regions where concave loops are left close to the surface. There are two distinct hardening mechanisms in this theory – the self hardening of the slip line due to the generation of debris on the slip plane itself, and the interaction hardening of one slip line by the debris produced in a neighbouring slip line. While the theory has certain attractions in that it explains the hardening rate in terms of a one-parameter dislocation distribution, there are, as always, difficulties. Firstly, the theory assumes the relation between debris dislocation density and flow stress given by (5.11) but the mechanism for this flow stress relation is not at all clear. Actually the loop debris will effectively reduce the spacing between obstacles in the slip plane as well as generating an internal stress, rather like the case of Orowan loops around particles in dispersion-hardened alloys (see above). Such a mechanism will give rise to hardening which no doubt can be put into the form of (5.11), but the constant $\alpha$ will have a value appropriate to this mechanism and has yet to be calculated. Secondly, and related to the first point, it has not been established what happens when a second dislocation passes on the same plane or on one close to that of the first. It would be expected that some of the concave loops formed previously would collapse so that the net rate of formation of debris is likely to decrease. In the extreme case, if after a few dislocations have passed the debris density on a slip plane remained constant no further work hardening need take place because the slip line spacing can always adjust itself so that the interaction hardening is zero; the theory should therefore be extended to consider the self hardening of a slip line when several dislocations have passed before a work hardening theory is developed. Kocks suggests that the dislocation loops are stabilised by secondary slip but this proposal would have to be considered in detail, and the problem now reduces to that encountered in all 'pile-up' theories, i.e. the problem of plastic relaxation. Thirdly, the interaction between glide dislocations on neighbouring planes is ignored completely. This is because Kocks considers only the self hardening of a slip line in the absence of dislocations on parallel slip planes. This seems a rather drastic assumption which needs justification; the implication is that dipoles or multipoles are not formed, or if they are, do not affect the work hardening rate. The neglect of interaction between gliding dislocations is also consistent with Kocks's view that the formation of linear obstacles in the slip plane is unlikely. Fourthly, in applying the theory to experiment, the significance of point obstacles is not clear; the electron microscope observations at the end of stage I show the presence in the slip plane of linear obstacles which are hard to reconcile with the Kocks model. It is also not clear how the model could generate the sheet- and wall-like structures which are formed in stage II (see e.g. Steeds, 1966). Finally, in the model slip lines would be expected to grow gradually with increasing stress whereas the experiments show that slip lines are fully formed in a small interval of stress and then stop. Kocks suggests that sources are stopped by the back stress from dislocation loops which have become trapped at local regions of high obstacle density resulting from processes on neighbouring planes. Such loops will, however, generate stresses elsewhere in the crystal, and these are not taken into account.

It seems unlikely that Kocks's model can form a good basis for work hardening in stage II, although the process of formation of concave loops around local hard regions is likely to occur. The theory is of course more relevant to hardening in dispersion-hardened alloys where, however, the debris, initially in the form of concave loops, is formed at every obstacle in the slip plane. Even then the neglect of interaction between glide dislocations seems questionable.

In the jog theory (proposed by Hirsch and reported by Mott, 1960) the flow stress is determined by the number of sessile jogs on the source dislocation, the jogs being produced by secondary slip. When the source operates the dislocations emitted have initially a low density of jogs, but as they travel through the crystal they accumulate jogs and are eventually stopped. The difficulty with this theory is that the dislocations acquire very nearly the necessary density of jogs in a distance of the order of the forest spacing, which is small compared with the slip line length. There is therefore no real physical criterion for the slip line length. Furthermore, the amount of secondary slip necessary to explain the observed hardening rate is about 30 per cent of the primary slip and this is rather large compared with that inferred from the X-ray data (Ahlers and Haasen, 1962; Mitchell and Thornton, 1964), although recently larger amounts of secondary slip have been deduced from shape-change measurements (Johnson,

Kocks and Chalmers, 1968). The jog theory would also predict a rather pronounced orientation dependence which is not observed.

*Long-range stress theories.* Theories of work hardening based on long-range stresses from piled-up groups have been developed by Mott (1952*b*), Friedel (1955), and Seeger, Diehl, Mader and Rebstock (1957). The basic idea in these theories is that dislocations are piled up in groups of $n$ at obstacles in their slip planes. The pile-ups generate long-range stresses which control the flow stress. In Mott's and Friedel's theories it is recognised that the large stresses close to the pile-ups will activate slip on secondary systems, thereby relaxing the long-range stress field. In Seeger's theory, plastic relaxation is neglected and secondary dislocations play essentially only a catalytic role in generating, by reacting with primary dislocations, the Lomer–Cottrell dislocations, which act as the dislocation obstacles. In the theory of Hirsch and Mitchell (1967, 1968) the problem of plastic relaxation is developed further; piled-up groups are transformed, by interaction with secondary dislocations, into complex boundaries. The density of forest dislocations is controlled by the internal stresses, and this leads to the view in this theory that the long-range stress and forest stress terms are proportional to one another.

Seeger's theory has been debated at length in the literature (e.g. Seeger, Diehl, Mader and Rebstock, 1957; Seeger, Mader and Kronmüller, 1963; Nabarro, Basinski and Holt, 1964; Seeger, 1968; Kronmüller, 1967; Hazzledine and Hirsch, 1967; Hirsch and Mitchell, 1967, 1968). Briefly, the flow stress $\tau_g$ is assumed to be given by the long-range stress due to the piled-up groups of $n$ dislocations, which to a first approximation can be treated as a super dislocation. The flow stress is defined as the maximum internal stress which has to be overcome in pushing a dislocation through the softest region of the crystal, i.e. half-way between neighbouring piled-up groups of $n$ dislocations, i.e.

$$\tau_g = \frac{\alpha n b G}{2\pi}\rho_p^{1/2} = \frac{\alpha n^{1/2} b G}{2\pi}\rho^{1/2} \qquad (5.40)$$

where $\rho_p$ is the density of piled-up groups, $\rho$ is the primary dislocation density, and $\alpha$ is a constant $\sim 1$. The density of pile-ups is related to strain by an equation analogous to (5.15), i.e.

$$\frac{d\rho_p}{d\epsilon} = \frac{1}{nLb} \qquad (5.41)$$

where $L$ is the slip line length. $n$ is taken as the maximum number of dislocations which can be pressed into a pile-up; the effective stress acting on the pile-up is taken as $\beta\tau$ where $\tau$ is the applied stress, and where $\frac{1}{2} < \beta < 1$, the parameter $\beta$ taking into account the screening due to other pile-ups. The relation is

$$\beta\tau = \frac{\pi}{4}\frac{nbG}{L}. \tag{5.42}$$

Since $\tau$ is put equal to $\tau_g$ these equations can now be used (assuming $n$ to be constant) to derive the work hardening rate which is found to be $\theta_{\mathrm{II}}/G = \alpha^2\beta/2\pi^3$. A more elaborate treatment (Seeger, Diehl, Mader and Rebstock, 1957) gives

$$\theta_{\mathrm{II}}/G = \frac{\alpha^2\beta}{6\pi^3}. \tag{5.43}$$

The experimental values of $\theta_{\mathrm{II}}/G$ fall within the range $\frac{1}{2} < \beta < 1$, as can be seen from fig. 5.22. Thus Seeger's theory is successful in explaining the observed work hardening rates; it also apparently explains the variation of slip line length with strain (equation(5.1)) through (5.42).

The basic assumption of Seeger's theory is that a pair of pile-ups of edge dislocations of the same sign, or a single screw pile-up, formed

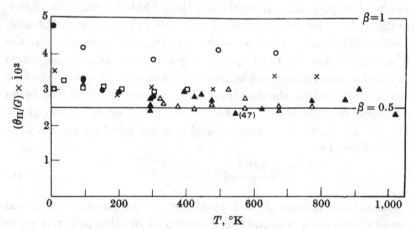

Fig. 5.22. Work hardening rate in stage II, $\theta_{\mathrm{II}}$, divided by the shear modulus, $G$, for various f.c.c. metals and alloys. ●, Al (see Berner, 1960; Noggle and Koehler, 1957); □, Cu (see Diehl and Berner, 1960; Köster, 1966); ○, Au (see Berner, 1960); ×, Ni (see Haasen, 1958; Meissner, 1959; Mader et al., 1963); △, Ni–50at.%Cu (see Pfaff, 1962); ▲, Ni–60at.%Co (see Pfaff, 1962)

at a stress $\beta\tau < \tau$, can generate a back stress equal to $\tau$ in the 'soft' region between the pile-ups. Hazzledine and Hirsch (1967) have shown that this is possible if the discrete nature of the dislocations is taken into account, but it turns out that the parameter $\alpha$ is rather smaller than that assumed by Seeger. Thus, if all the pile-ups are in the unfavourable configuration considered by Seeger it is found that for $\beta = \frac{1}{2}$ at most about 60 per cent of the flow stress could be due to the back stress from primary dislocations. The effective back stress is even less for the region between pile-ups of opposite sign. The electron micrographs suggest that this is a favoured arrangement (see figs. 5.7, 5.9).

Experiments on the hardening of secondary systems after slip on the primary system (latent hardening experiments) show that the hardening on systems coplanar with the primary is the same as that on the primary system, while it is considerably greater on secondary intersecting systems (Jackson and Basinski, 1967; Kocks and Brown, 1966; Nakada and Keh, 1966); these results cannot be reconciled with long-range stress from primary dislocations only, which would predict a flow stress on the coplanar system equal to half that on the primary system (Stroh, 1953). These experiments leave no doubt that secondary systems play an important part in controlling the flow stress; they do this partly by the forest intersection mechanism and partly by modifying the long-range internal stress pattern.

Consider now the operation of sources in Seeger's model. A source can operate if a single dislocation given out by the source can overcome the flow stress. (5.40). Once the applied stress is sufficient for this to happen, the source is assumed to continue to operate until the back stress from the fully formed slip line reaches $-\beta\tau$ at the source, $+\beta\tau$ being the effective applied stress acting at the source before the pile-up is formed. The slip line is assumed to be formed in a small stress interval, in accord with observations (Mader, 1957). But as soon as the source begins to emit dislocations the back stress from the new dislocations will make it impossible for the source to operate again unless the applied stress is raised; in fact the stress would have to be raised by $\beta\tau$ before the slip line is fully formed. The only way Seeger's model could work appears to be if as soon as the source begins to operate, the effective applied stress there increases or the internal stress decreases. The physical significance of the back stress equation is therefore not at all clear. It is also rather curious that while the effective applied stress on the pile-up and at the source is $\beta\tau$, it is equal to $\tau$ at the point where the single dislocation has to overcome the internal stress.

The difficulty with Seeger's back stress equation is an example of a general problem in work hardening theories, associated with calculations of the back stress from particular dislocations in the presence of many others whose stress fields should be taken into account and which reduce the back stress at large distances. Seeger's stage I theory suffers from the same difficulty (see above). Although the use of the back stress equation seems untenable, Seeger's theory has some desirable features. Firstly, he assumes an inhomogeneous dislocation distribution, with 'soft' regions between pile-ups and 'hard' regions where the pile-ups are stopped. This seems to be essential if the blocking of slip lines is to be explained. Secondly, the idea of working out the flow stress from groups of dislocations formed at the same stress in a consistent manner is good (although unfortunately the calculations of Kronmüller and Seeger, 1961 were not self consistent – see Hazzledine and Hirsch, 1967). Apart from the difficulty with the magnitude and symmetry of the internal stress, the theory lacks any explicit relation for the mean free path of the dislocations in terms of obstacle density, and no attempt is made to explain the large observed density of forest dislocations, which affect the internal stress pattern.

The theory of Hirsch and Mitchell attempts to derive the dislocation arrangement as well as the work hardening rate. The basic idea is that at the end of stage I long barriers are formed consisting of bands of dipoles of primary dislocations stabilised and connected by secondary dislocations. In the transition region primary dislocations begin to pile up and more secondary dislocations are generated by the action of the internal stresses, until the region in which secondary slip occurs extends throughout the crystal. A structure characteristic of stage II is formed in which any newly formed slip line is converted into a complex 'mat' of high dislocation density by interaction with secondary dislocations. Each of these regions acts as a new obstacle to further slip. The secondary dislocations are generated by the action of the internal stresses from the original pile-ups, and the resolved component of the applied stress; these secondaries stabilise the primary arrays by elastic interaction.

Around the tip of each piled-up group there will be a region, extending from one pile-up to the next in the model, in which the maximum secondary slip has occurred compatible with the stresses

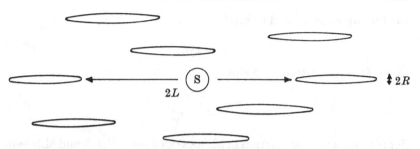

Fig. 5.23. Structure of elongated obstacles, representing dislocation 'mats' parallel to primary slip planes, in stage II; $R$ is the radius of interaction of an obstacle normal to the slip plane; $2L$ is slip length; S the position of a source. (Hirsch and Mitchell, 1967.) Reproduced by permission of the National Research Council of Canada from *The Canadian Journal of Physics* **45**, 663–706, 1967

acting on the secondary system. Each of these 'obstacles', in the form of flat ribbons extending roughly over the pile-up length parallel to the slip plane, is characterised by a radius of interaction $R$ defined by the condition that any new group of dislocations on a parallel plane within $2R$ of the pile-up will be stopped. The new dislocations will be stopped by elastic interaction with both primary and secondary dislocations in the obstacle. Fig. 5.23 shows schematically the structure assumed. As in Seeger's theory, the structure is inhomogeneous; within $2R$ of the obstacle the stresses from the primary and forest dislocations are sufficient to hold up a pile-up; the region between obstacles on neighbouring slip lines will be 'soft' and control the flow stress. A basic assumption in the theory is that the structure remains similar in stage II, but simply reduces in scale. This is the 'law of similarity' and is common to all stage II theories (Nabarro, Basinski and Holt, 1964; Kuhlmann-Wilsdorf, 1962, 1968).

The flow stress is assumed to be given by

$$\tau = \tau_0 + \tau_i + \tau_f \qquad (5.44)$$

where $\tau_0$ is the fraction stress, $\tau_i$ the long-range back stress from the dislocation mats and $\tau_f$ the forest contribution (this could also include the Frank–Read source term). $\tau_i$ and $\tau_f$ are assumed to remain in constant proportion because the structure remains similar; hence, since $\tau_i \propto 1/D$, where $D$ is the mean spacing between obstacles, we write

$$\tau - \tau_0 = k_3/D \qquad (5.45)$$

where $k_3$ is a constant. The radius of interaction $R$ will be related to the pile-up length $a$ and is written

$$R = k_1 a \qquad (5.46)$$

where $k_1$ is a constant, and since

$$a = \frac{nbG}{\pi(\tau - \tau_0)} \qquad (5.47)$$

(for screws: a similar relation holds for edges – see Hirsch and Mitchell, 1967) it follows from (5.45) and (5.46) that

$$R = k_4 D \qquad (5.48)$$

and also

$$D = k_2 a = (k_1/k_4) a \qquad (5.49)$$

where $k_2$, $k_4$ are other constants. The condition for blocking the slip lines (length $2L$) is

$$4\rho_p RL = 1 \qquad (5.50)$$

where $\rho_p = D^{-2}$ is the density of obstacles, equal to the density of pile-ups formed. Using (5.41) and (5.46)–(5.50), we find $\theta_{II}/G = 2k_1/\pi$; a more elaborate treatment (Hirsch and Mitchell, 1967) gives

$$\theta_{II}/G = k_1/3\pi. \qquad (5.51)$$

The work hardening rate therefore depends on the ratio of the obstacle radius to the pile-up length. Now the back stress at the back of the pile-up will act as an effective obstacle to groups of dislocations of the same sign; dislocation groups of opposite sign are likely to be trapped by the high forest density at the tip, or will form multipoles; the various arrangements expected are shown schematically in fig. 5.24.

So far, calculations for $k_1$ have been carried out only for groups of edge dislocations trapped by pile-ups with the same number of primary edge dislocations of the same or opposite sign, unmodified by secondary dislocations (Hazzledine, 1968, 1971a), corresponding to figs. 5.24(a), (b). Pande (1970) has carried out similar calculations assuming a continuum model. Hazzledine finds that as the distance between slip planes is decreased within a certain range, the fraction of the number of dislocations held up increases from 0 to 1. If all the dislocations are trapped, for groups of 20 dislocations of the same sign $k_1 \sim 1/50$; for groups of opposite sign $k_1 \sim 1/160$. The correspond-

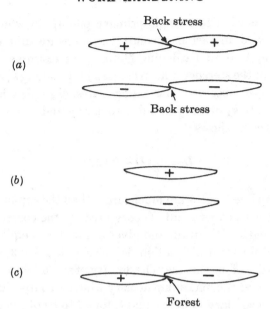

Fig. 5.24. Dislocation structures expected if newly-formed slip lines are held up (a) by back stress from previously formed pile-ups of same sign, (b) by forming multipoles with previously formed pile-ups of opposite sign, (c) by interaction with high density of forest dislocations near tip of previously formed pile-ups. (Hirsch and Mitchell, 1967.) Reproduced by permission of the National Research Council of Canada from *The Canadian Journal of Physics* **45**, 663–706, 1967

ing work hardening rates are $\theta_{II}/G \sim 1/470$ and $\sim 1/1500$; the former is of the correct order of magnitude. Both estimates will of course be modified by the presence of forest dislocations; furthermore, as the new group is being formed, secondary dislocations will be generated as the first few dislocations are being held up, so that the effective trapping radius is likely to be considerably larger, possibly corresponding to the limit when one dislocation is held up; the values of $k_1$ for this case are $1/14$, and $1/20$ respectively, and $\theta_{II}/G \sim 1/130$ and $\sim 1/190$. Thus the values of $k_1$ estimated in this way cover the range of observed work hardening rates.

Only one of the other three constants $k_2$, $k_3$, $k_4$ in the theory is independent; Hirsch and Mitchell use the criterion that secondary slip must occur throughout the crystal so that the total stress on secondary systems must be sufficient to propagate slip across the primary slip bands; they argue that the flow stress on secondary systems must be exceeded in the region between neighbouring pile-ups; this defines $k_2$. Estimates for $k_2$ are made from the extent

of the region, normal to a single primary pile-up, in which the flow
stress is exceeded on various secondary systems for different crystal
orientations and latent hardening factors. For example, for copper
in the middle of the stereographic triangle $k_2 \sim \frac{1}{2}$; taking $k_1 \sim \frac{1}{30}$, which
would give the observed hardening rates, $k_4 \sim \frac{1}{15}$ (Hirsch and
Mitchell estimate $k_4$ directly in a different way and find $k_4 \sim \frac{1}{8}$). The
detailed treatment shows

$$L = \frac{3}{2k_4} D \sim 22.5D \tag{5.52}$$

so that the slip line lengths are much larger than the slip line spacings,
in agreement with experiment. Theory predicts the correct variation
of slip line length with strain and gives a flow stress equation of the
type of (5.10) for total dislocation density, with $\alpha \sim 0.3$, which is of
the correct order of magnitude. The density of secondary dislocations,
$\rho_s$, is calculated as the maximum density which is compatible with the
stress on the secondary systems, and is found to be of the same order
as the primary dislocation density, $\rho_p$, in agreement with experiment.
The *minimum* amount of plastic strain on the secondary system is
$\sim D/2L \times \rho_s/\rho_p$ times the strain on the primary system, and this
ratio comes out to be only $\sim 2$ per cent, which is also compatible with
experiment. The basic dislocation structures predicted from the model
and illustrated in fig. 5.24 are all observed; these structures possess a
pronounced anisotropy which is consistent with latent hardening of
secondary systems (Jackson and Basinski, 1967).

The theory predicts that for solution-hardened alloys, the presence
of the friction stress leads to a smaller fraction of secondary to primary
dislocation densities, and to a somewhat smaller work hardening
rate; these results are in general agreement with experiment (Hirsch
and Mitchell, 1967, 1968) but the dislocation arrangements in such
alloys appear to be much more uniform than expected from the
theory, so that the applicability of the latter is not clear (Pande and
Hazzledine, 1971b).

The formation of slip lines with $n$ dislocations in a small stress
interval is explained in terms of a model in which sources act co-
operatively, the internal stress generated by one group aiding the
operation of the source of the other. The model is illustrated in fig.
5.25, where successive dislocations from one source initiate successive
dislocations from the other source. The process stops when the back
stress at the source due to one set of dislocations is equal to the internal

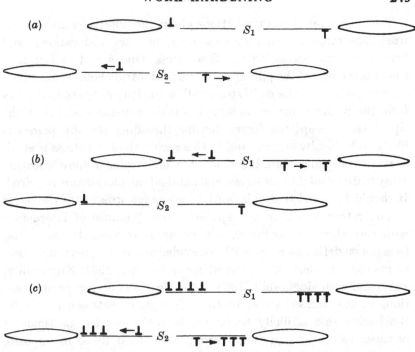

Fig. 5.25. Co-operative operation of sources. Successive dislocations from one source initiate successive dislocations from the other source. (Hirsch and Mitchell, 1967.) Reproduced by permission of the National Research Council of Canada from *The Canadian Journal of Physics* 45, 663–706, 1967

stress generated by the other. Locally, the piled-up group from one source is estimated to generate a maximum of about half the applied stress at the other source. Thus increasing the applied stress by a small amount, $\Delta\tau$, 'triggers' a co-operative mechanism in which many more dislocations can be generated than would be expected if their number were controlled by their back stress, in the absence of the co-operating dislocations. The theory predicts an orientation dependence of the various constants $k_i$; Hirsch and Mitchell (1967, 1968) have estimated the variation of $k_2$ with orientation and, assuming that $k_4$ is independent of orientation, derive the orientation dependence of the work hardening rate, which turns out to be quite small, in agreement with observation. The reason for this is that the internal stresses generated by the primary dislocations play a very important part in controlling the slip on the secondary systems. However, the assumption that $k_4$ is orientation independent is very doubtful, and detailed calculations are required.

The theory makes no predictions about the relative contributions from long-range internal stresses from primary dislocations, and from the forest stress, to the flow stress. One would expect that towards the rear of the pile-ups or mats (dislocation boundaries which are not stress-free) the main stress will be the long-range contribution from the primary and secondary dislocations in the mats; near the tip of the pile-ups the forest density threading the slip planes is likely to be locally largest, and in this region the back stress is small or may even change sign. It is not clear which region in the dislocation array will control the flow stress, and detailed calculations are required. It should be noted that substantial long-range stresses, of the order of but rather less than the applied stress (because of interaction with secondary dislocations), will occur at or close to the mats; thus the model is consistent with the evidence from magnetic measurements (Kronmüller, 1959; Kronmüller and Seeger, 1961; Kronmüller, 1967) and with Mughrabi's (1968, 1971) electron microscope observations in the stressed state. In the soft regions, between the mats, the back stress is likely to be considerably smaller, particularly because dislocations in neighbouring mats tend to be of opposite sign. Thus it seems possible that the flow stress may be controlled by forest dislocations near the tip of the pile-up; this would be consistent with the important results of Jackson and Basinski (1967), who showed that when a crystal is deformed on a second system after deformation on the first system, the temperature and strain rate sensitivity of the flow stress remain the same, independent of the dislocation distribution – a result most easily understood if forest dislocations are controlling the temperature dependent and temperature independent parts of the flow stress (see section 5.5). Jackson and Basinski also find that the flow stress for slip on the secondary system varies linearly (except for deviations in early stage I) with the flow stress on the primary system; they show that this result is explained on the Hirsch–Mitchell model provided the flow stress is always determined by the forest dislocations; the result implies that the density of forest dislocations is always controlled by the internal stresses generated by the primary dislocations, and that the applied stress component on secondary systems has a relatively small effect.

One of the important features of the model is that it explains the fact that the densities of primary and secondary dislocations produced in stage II are comparable. Essmann (1965a, b) showed that the density of primary dislocations in copper varies linearly with stress in stages I and II, while that of secondary dislocations varies as (stress)$^2$. Seeger (1968) argues that this result contradicts the assumption of the 'law of similarity' made in the model. The significant difference in the variation of the two types of dislocation density with strain occurs, however, in stage I and early stage II where of course the very large density of multipoles on the primary system, formed in stage I and in the transition region, and which has little effect on stage II hardening, constitutes an important part of the primary dislocation density. It should be mentioned that Seeger's theory depends as much as all others on the 'law of similarity' since he uses a flow stress equation with a constant $\alpha$ independent of strain. This equation also predicts that the *primary* dislocation density is proportional to the (stress)$^2$. It is now clear, and in agreement with views expressed by Seeger (1968), that in any long-range stress theory the stress from the complex boundaries found by transmission microscopy must be calculated, i.e. the effect of secondary dislocations cannot be neglected.

In conclusion, it appears that the model of piled-up groups modified by secondary slip to form complex boundaries, which are not stress free, is capable of explaining in a qualitative and to some extent semiquantitative way, many of the experimental results on hardening dislocation distribution, and slip lines. This model is essentially a development of Mott's original model (1952b). The proper test of the theory must, however, await further self-consistent calculations of the various parameters $k_i$ in the theory, of the internal stresses from the boundaries, and of the forest stress.

Finally we should like to note the basic differences between stages I and II. In stage I the internal stresses from the primary dislocations are largely cancelled out by formation of multipoles. The residual long-range stresses are small and lead to little work hardening and a rate of multiplication of forest dislocations much smaller than that of primary dislocations. The multipoles themselves are very weak obstacles to primary slip. In metals in which cross-slip is relatively easy, screw dislocations tend to annihilate by cross-slip, induced by the internal stresses from screws of opposite sign (Steeds and Hazzledine, 1964; Essmann, 1964; Hirsch and Lally, 1965); this process is inhibited in alloys of low stacking fault energy and/or large friction stress (Taylor and Christian, 1967b). In stage II piled-up groups are formed at strong obstacles; the internal stress due to these groups generates secondary dislocations, comparable in density to that of the primary dislocations, which interact with one another forming stable dislocation boundaries which are not stress-free and which act as strong obstacles. The long-range stress from any one dislocation boundary extends only over a region of the order of the boundary spacing, since neighbouring boundaries tend to be of opposite sign. There will be substantial long-range stresses from the complex boundaries in the region between them, particularly close to the boundaries. Screw dislocations will be prevented from mutual annihilations by interaction with secondary dislocations. The flow stress has contributions from long-range stress (from the mats) and forest terms and there is doubt about their relative magnitudes.

## Acknowledgment

The author would like to acknowledge the numerous stimulating discussions on this complex topic with, and the encouragement and active support of Sir Nevill Mott during the author's period in the Cavendish Laboratory, Cambridge.

# CHAPTER 6

# FRACTURE

## by A. H. COTTRELL†

## 6.1 ATOMIC FORCES AND THE STRENGTH OF SOLIDS

Professor Sir Nevill Mott was one of the first to realise, at the beginning of the 1930s, that, with the arrival of quantum mechanics, the road at last lay open for the creation of a general theory of bulk matter in terms of atomic and electronic structure and properties. The great programme of solid state physics was thus begun.

In parts of this programme, for example in the theory of metallic conductivity, a quantum theory of solids could be developed directly by finding suitable, if approximate, solutions of Schrödinger's equation. In others, however, the physical situation proved too complex for this approach. The strength and plastic properties of solids, for example, are not determined by the average behaviour of the atoms but by the exceptional behaviour of those relatively few atoms situated at lattice irregularities. The theories of such properties could not be developed without providing some intermediate concepts, to by-pass the mathematically formidable and to some extent physically irrelevant problem of solving Schrödinger's equation for lattice irregularities, and to enable the theory to work directly in terms of the atoms and their movements. One of the most useful of these has been the representation of the cohesion of a solid in terms of 'atomic bonds', i.e. force–displacement relations between pairs of atoms such as that in fig. 6.1.

For covalent solids this representation can be justified quantum-mechanically, but for metals it is highly artificial and must be used with care. Such force–displacement relations are familiar, both from the theory of molecules and from the theory of the ideal strength of solids. Molecular theory shows that, commonly, the maximum interatomic attraction $f_m$ is reached at a spacing about one-fifth

---

† Sir Alan Cottrell is the Master of Jesus College, Cambridge. He was formerly the Chief Scientific Adviser to H.M. Government.

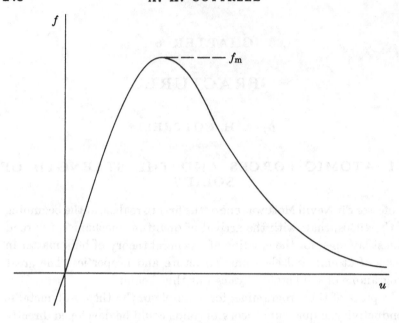

Fig. 6.1. Force $f$ between atoms as a function of displacement $u$ from the equilibrium spacing

larger than the equilibrium spacing. In solids, the slope $df/du$ at the equilibrium spacing is responsible for the elastic constant, i.e. Young's modulus $E$ for simple tension or the shear modulus $\mu$ for shear. It follows that the order of magnitude of the ideal strength $\sigma_m$ is

$$\sigma_m \simeq 0.1E \quad \text{(tension)}$$
$$\simeq 0.1\mu \quad \text{(shear)}. \tag{6.1}$$

Most of the work of failure is done after the maximum $f_m$ has been passed. Because of this important instability in the law of force, the process of mechanical failure in a solid concentrates itself in some highly localised region. Such a region may exist *ab initio*, as an original defect, or it may form when the applied stress approaches $\sigma_m$, for example by a chance thermal fluctuation. In either case the failure then propagates outwards from this origin, often along a plane of maximum tensile or shear stress in the material. There are three basic types of failure. First, crystallographic slip, in which the atomic bonding is perfectly restored in the slipped region, and for which the characteristic fault is the unit dislocation with a constant Burgers vector. Second, tensile cracking, in which the separation of the surfaces of failure precludes any restoration of broken bonds.

The characteristic fault in this case, i.e. the crack itself, can be regarded as a dislocation of variable Burgers vector. Third, shear faulting such as occurs in compressive fractures in rock and other brittle materials. This is an imperfect form of slip in which, through branching of the fault plane and creation of debris between the slipping surfaces, the bonding is not restored in the fault.

## 6.2 DISLOCATIONS, CRACKS AND FAULTS

Volterra's theory (1907) of dislocations in an elastic continuum enables us to describe all these modes of mechanical failure in a unified way (see also chapter 2). For simplicity we take a cylindrical body, as in fig. 6.2, cut it along its negative $x$-axis as far as the centre line ($z$-axis) and then, by applied forces, create a displacement field $\mathbf{u}\ (= u, v, w)$ in it. This field in general separates pairs of points such as

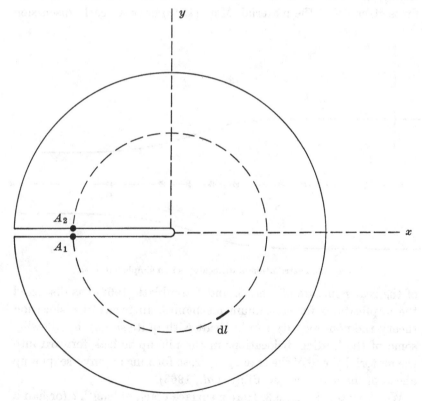

Fig. 6.2. Volterra dislocation

$A_1$ and $A_2$, initially coincident on the cut faces, by a Burgers vector
$\mathbf{b}$ ($=b_1$, $b_2$, $b_3$), where

$$\mathbf{b} = \int \frac{\partial \mathbf{u}}{\partial l} \, dl \tag{6.2}$$

this being the integral of increments of elastic displacement over
elements $dl$ of a large circuit from $A_1$ to $A_2$ round the end of the cut.

Cracks and shear faults are represented by the condition that the
distribution of force between the cut faces is constant over those
faces (zero for a free crack). The fact that $\mathbf{b}$ is then not a constant
is easily dealt with by regarding $\mathbf{b}$ as a linear distribution of infinitesimal unit dislocations, as in some theories of crystal slip bands.

The piling-up of such dislocations in an infinitely sharp crack
produces an infinite stress at the tip of the crack. This unreal result is
avoided by abandoning linear continuum mechanics at the tip and
taking account, instead, of the discrete atomic structure and the
finite strength of the material. Mott (1948) gave an early discussion

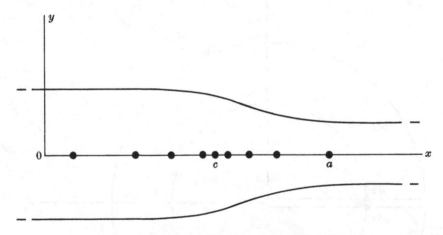

Fig. 6.3. Distribution of dislocations in a simple half-crack

of the real structure of a crack and Barenblatt (1962) has discussed
the implications for continuum mechanics. In terms of dislocation
theory the non-linearity can be dealt with quite simply by allowing
some of the leading dislocations in the pile-up to leak forward into
the material ahead of the crack proper, so forming an 'inverse' pile-up
ahead of the main one (cf. Bilby *et al.*, 1963).

We show this in fig. 6.3. Here a surface crack of length $c$ (or half a
symmetrical internal crack of length $2c$) is represented by a sequence

of dislocations which extends beyond the crack proper into the 'preparatory zone' $c < x < a$. Let $b'(x)\mathrm{d}(x)$ be the total length of Burgers vector for all those dislocations between $x$ and $x + \mathrm{d}x$. The relative displacement of the crack faces at $x$ is then given by

$$b(x) = \int_x^\infty b'(x)\,\mathrm{d}x. \tag{6.3}$$

In a state of equilibrium, the total stress acting on each dislocation is zero. This total stress is composed of the nominal applied stress $\sigma$, the resistance stress due to the strength of the material, $\sigma_i(b)$, which opposes the passage of a dislocation, and the combined stresses of all the other dislocations, as given by the usual formulae of dislocation theory. The equilibrium condition then becomes

$$\sigma - \sigma_i = A \int_{-\infty}^{+\infty} \frac{b'(x')\,\mathrm{d}x'}{x' - x} \tag{6.4}$$

where

$$A = \mu/2\pi k \tag{6.5}$$

and $k$ ranges from $1 - \nu$ for edge dislocations to $1$ for screw dislocations, $\mu$ and $\nu$ being the shear modulus and Poisson's ratio, respectively.

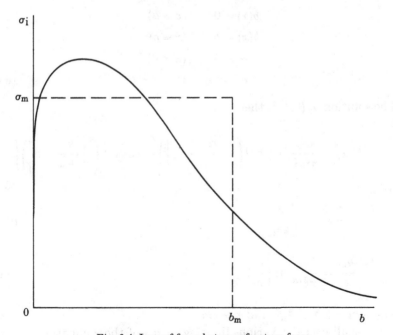

Fig. 6.4. Law of force between fracture faces

It is convenient, in describing the law of force $\sigma_i(b)$ between the fracture faces, to take out the linear elastic component of the total deformation $b$ and describe instead the relation between $\sigma_i$ and the non-Hookeian deformation associated with it. The curve in fig. 6.4 shows a typical relation. This might represent, for example, an atomic force–displacement curve (cf. fig. 6.1) as in a simple brittle tensile fracture; or a work hardening and tensile necking curve in a ductile plastic fracture; or the deformation and failure of the adhesive in a glued joint.

Equation (6.4) is not easily solved for such general laws of force, however. As well as mathematical complexity there is the more fundamental difficulty that neither the crack nor its preparatory zone are well defined. The lack of a definite Hookeian elastic limit means that $a \to \infty$; and an asymptotic approach of $\sigma_i$ to zero at large $b$ means that $c \to 0$.

To avoid these difficulties we use a simple rectilinear approximation to the law of force, as shown in fig. 6.4, in which the material obeys Hooke's law perfectly up to a limiting strength $\sigma_m$ and then deforms at constant stress to a limiting displacement $b_m$ beyond which it totally fails (Bilby et al., 1963). We thus have, in fig. 6.4,

$$
\begin{aligned}
b(x) &= 0 & (x > a) \\
b(x) &= b_m & (x = c) \\
\sigma_i &= 0 & (x < 0) \\
\sigma_i &= \sigma_m & (c < x < a).
\end{aligned}
\tag{6.6}
$$

The solution of (6.4) is then

$$
b'(x) = \frac{\sigma_m}{\pi^2 A} \left[ \cosh^{-1}\left( \left| \frac{m}{c-x} + n \right| \right) - \cosh^{-1}\left( \left| \frac{m}{c+x} + n \right| \right) \right]
\tag{6.7}
$$

$$
\frac{c}{a} = \cos\left( \frac{\pi\sigma}{2\sigma_m} \right)
\tag{6.8}
$$

$$
\frac{b_m}{c} = \frac{2\sigma_m}{\pi^2 A} \ln\left( \frac{a}{c} \right)
\tag{6.9}
$$

where $m = (a^2 - c^2)/a$ and $n = c/a$.

We shall consider various limiting forms of these equations.

## 6.3 CLASSICAL CRACKS

The extreme limit $\sigma_m \to \infty$ and $a \to c$ gives the classical elastic theory of an infinitely sharp slit in an infinitely strong medium (e.g. Sneddon, 1946). The slit is deformed elliptically, i.e.

$$b(x) = b_0 \frac{(c^2 - x^2)^{1/2}}{c} \quad (x < c) \tag{6.10}$$

with

$$b_0 = \frac{\sigma c}{\pi A} \tag{6.11}$$

and the stress varies as

$$\sigma(x) = \sigma \frac{x}{(x^2 - c^2)^{1/2}} \quad (x > c) \tag{6.12}$$

along the $x$-axis.

## 6.4 GRIFFITH CRACKS

A more realistic case is that of a medium with limited strength under a small applied stress, i.e. $\sigma \ll \sigma_m$. If $R = a - c$ we have $R \ll c$ and (6.8) and (6.9) reduce to

$$\frac{R}{c} = \frac{1}{2} \left( \frac{\pi \sigma}{2\sigma_m} \right)^2 \tag{6.13}$$

$$\frac{b_m}{c} = \frac{2\sigma_m}{\pi^2 A} \frac{R}{c}. \tag{6.14}$$

We see from the second of these that, in this limit where $\sigma \ll \sigma_m$, $R$ is a true property of the material, $R_0$, independent of $\sigma$ and $c$; thus

$$R_0 = \frac{\pi \mu b_m}{4(1 - \nu)\sigma_m}. \tag{6.15}$$

This defines the 'width', $a - c$, of the crack front, just as the very similar formula

$$w \simeq \frac{\mu b}{2\pi(1 - \nu)\sigma_m} \tag{6.16}$$

defines the width of a crystal dislocation (Foreman et al., 1951). In both cases the width varies inversely with limiting strength.

Since $R = R_0 =$ constant in (6.13) it follows that the equilibrium applied stress, $\sigma$, varies as $c^{-1/2}$. This fundamental result, first derived by Griffith (1921) for a special form of crack, has a general validity at low stresses, irrespective of the detailed atomic processes that govern the law of force $\sigma_1(b)$. Equations (6.13) and (6.14) give the *Griffith stress* as

$$\sigma = \left(\frac{4A\sigma_m b_m}{c}\right)^{1/2} = \left(\frac{8A\gamma}{c}\right)^{1/2} \tag{6.17}$$

where $2\gamma$ is the *work of fracture*, i.e.

$$2\gamma = \int_0^\infty \sigma_1(b)\, db \tag{6.18}$$

and for the rectilinear law of force

$$2\gamma = \sigma_m b_m. \tag{6.19}$$

In the limiting case of totally brittle fracture $\gamma$ is the specific surface energy of the fracture faces.

In his original derivation of equation (6.17) Griffith balanced the release of elastic energy against the increase of surface energy, during an incremental growth of the crack. The total energy of a plane strain crack of length $2c$ is given by

$$W = -\frac{\sigma^2 c^2}{4A} + 4\gamma c \tag{6.20}$$

and the equilibrium condition $dW/dc = 0$ then gives (6.17). Since the release of elastic energy increases as $c^2$ whereas the surface energy increases as $c$, the equilibrium is unstable under a constant applied stress $\sigma$. For this reason the Griffith stress can often be interpreted as the nominal fracture stress of a cracked solid.

There are several limitations to the above analysis. First, it is restricted to the range $\sigma \ll \sigma_m$; we shall discuss the behaviour at higher stresses later. Second, the analysis is based on a uniform nominal applied stress, at least over the neighbourhood of the crack. Other stress distributions give different results. For example, if a crack is sprung open by a fixed wedge inserted between its faces, the equilibrium is stable (Mott, 1948; Benbow and Roesler, 1957). The crack grows to a fixed length, governed by the thickness $b_0$ of the insert and the elastic constants and surface energy. The measurement of stable crack lengths in fact provides a means for determining the surface energies of brittle materials such as mica (Obreimow, 1930; Bailey, 1957). The equilibrium length of a surface crack can be found most simply from the Foreman–Nabarro dislocation theorem (cf. Cottrell, 1964). The misfit energy created by a fixed insert of thickness $b_0$ is $\mu b_0^2/4\pi(1 - \nu)$. Equating this to $2\gamma c$, we have

$$c = \frac{\mu b_0^2}{8\pi(1 - \nu)\gamma}.$$ (6.21)

If $b_0 = 1$ atomic spacing and $8\gamma = \mu b_0$ we obtain $c \simeq 0.5b_0$; the theory is of course highly inaccurate in this limiting case.

An example in which the loading conditions make Griffith's theory quantitatively inapplicable is that of a longitudinal slit in a cylindrical pressure drum. The sides of the drum along the edges of the slit bulge out under the pressure, which creates bending and membrane stresses so that the slit appears to be subjected to a uniform nominal stress that increases with its length (Wells, 1962; Irvine, Quirk and Bevitt, 1964).

The Griffith equation is based on an energy balance. It is thus a general thermodynamic condition, necessary but not sufficient for the growth of a crack. The additional condition to be met is that there must be an operable mechanism of fracture at the tip of the crack. Consider, for example, the question of starting a fracture from a blunt notch. The notch may be deep enough to satisfy the Griffith condition but too blunt to raise the applied stress àt its tip to the fracture strength of the material. As a more extreme example, consider a sharp notch loaded up to its Griffith stress, but loaded in compression.

People have sometimes failed to separate these quite distinct thermodynamic and mechanistic conditions for fracture, mainly because under some circumstances the thermodynamic condition is sufficient. Consider, for example, a completely brittle tensile crack. The very structure of this is such as to ensure that there is always a sequence of atomic bonds, at all stages of elongation and failure, in its preparatory zone; cf. fig. 6.6(b). The Griffith condition may also be sufficient in some chemically or thermally assisted fractures in which atoms are removed from the tip of a notch by dissolution or evaporation. The notch could then be made to grow deeper by any applied stress, tensile or compressive, that exceeded the Griffith stress.

## 6.5 FRACTURE AT HIGHER STRESSES: NOTCH SENSITIVITY

Shorter cracks have higher equilibrium stresses. By expanding (6.8) as far as $(\sigma/\sigma_m)^2$ and substituting in (6.9) we can obtain a Griffith equation again but with $c$ replaced by $c[1 + (\pi\sigma^2/24\sigma_m^2)]$. This increase in effective crack length, first deduced by Irwin (1960), follows from the fact that $R$ is no longer negligible compared with $c$. Because the material in the preparatory zone is partially failed, the effective length of the crack is increased by an amount equivalent to a total failure over a length between $c$ and $c + R$.

At high $\sigma/\sigma_m$ values a Griffith approximation is no longer possible. Equation (6.17) is then replaced by the general form

$$\sigma = \frac{2\sigma_m}{\pi} \cos^{-1}\left[ \exp\left( -\frac{\pi^2 A b_m}{2\sigma_m c} \right) \right] \tag{6.22}$$

in which $\sigma \to \sigma_m$ as $c \to 0$. The range of crack lengths above which Griffith's formula is valid and below which the fracture strength $\sigma$ approximates to the limiting value $\sigma_m$ is broadly indicated by

$$c = R_0 \tag{6.23}$$

at which $\sigma = 0.7\,\sigma_m$. This gives a criterion for notch sensitivity.

## 6.6 TENSILE FRACTURE

We now apply our general considerations to tensile cracks, as in fig. 6.5, strained in plane deformation by a nominal tensile stress $\sigma$ directed along the $y$-axis. We regard the material above and below such a crack as two large, linearly elastic, half-bodies and these are joined together across the plane of the crack itself by a thin adhesive layer, the non-linear stretching and failure of which is represented by edge dislocations.

The crack grows by the continuous creation of these dislocation lines, in pairs along the $z$-axis, and by the symmetrical separation, outwards along the $x$-axis, of the members of each pair. This is a brittle fracture in the sense that the dislocations remain localised in the adhesive layer so that, when completely separated by the crack, the two elastic half-bodies completely recover their undistorted forms. This macroscopically brittle failure can be produced by many types of microscopic failure within the adhesive layer – brittle, ductile, viscous, etc. – all of which can be described formally by this same dislocation mode.

This dislocation mode is *cumulative* and *dilatational*. By cumulative we mean that the same group of dislocations, sufficient to start the fracture locally, is also sufficient to complete the whole fracture by

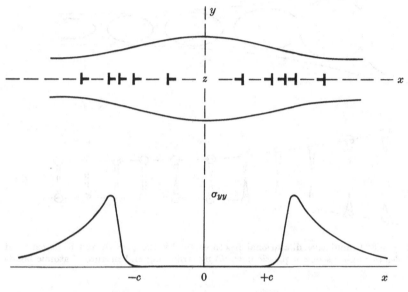

Fig. 6.5. Plane strain tensile fracture

running along the (uniform) adhesive layer to the ends of the $x$-axis. The additional dislocations created behind them by the opening-up of the crack merely help the applied stress to push this leading group along the layer. The leading group constitutes the preparatory zone, which runs along with the group ahead of the crack proper. This cumulative mode gives an unstable fracture, under a constant applied stress, since the additional dislocations help the applied stress to spread the crack.

The dilatational character of the fracture, caused by the separation of the crack faces, is represented by the climb of edge dislocations along the adhesive layer. In an ideally brittle fracture this dilatation is produced by the stretching of the atomic bonds which join the two half-bodies. However, when the adhesive layer fails by a process of flow, e.g. plastic or viscous, the dilatation must occur by the growth of holes as, for example, in fig. 6.6(*a*). Such a *discontinuous* mode of cracking, produced by the growth and coalescence of holes in the preparatory zone, is thus an essential feature of macroscopically brittle tensile fractures produced by non-dilatational processes of failure at the microscopic level.

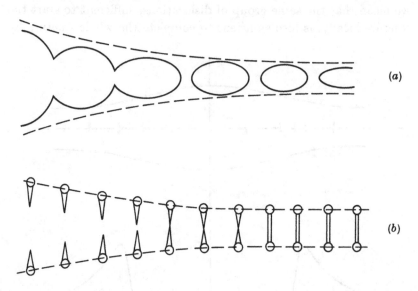

Fig. 6.6. Cumulative dilatational fractures (a) by the growth and coalescence of holes through viscous or plastic flow, (b) by stretching and rupture of atomic bonds

## 6.7 IDEALLY BRITTLE TENSILE FRACTURE

We now consider microscopically brittle tensile fracture, such as ideal cleavage in crystals. The adhesive is now merely the sheet of atomic bonds which join the two elastic half-bodies, as in fig. 6.6(b). If this particular mode of failure is weaker than any other, a question we shall take up later, Griffith's criterion is both necessary and sufficient for determining the fracture stress because, as we have seen, the geometry of the crack front necessarily provides an atomic mechanism of failure. For the crack to advance by one atomic spacing, each atomic bond in the preparatory zone has simply to increase its elongation to that of its predecessor.

This is formally similar to the process of slip in the plane ahead of a moving glide dislocation. The width of the crack front, when $\sigma_m \simeq \mu/4 \simeq E/10$ and $b = 1$ atomic spacing, is about 4 atomic spacings; cf. equation (6.15). In the language of dislocation theory the crack front is thus fairly 'wide' and, from the theory, we expect its 'Peierls–

Nabarro force' to be extremely small. It then follows, again from the theory, that ideally brittle crack propagation is virtually an athermal process. There is nothing to prevent such a crack, in a perfect crystal, acquiring a speed approaching that of elastic waves.

Mott (1948) recognised that the limiting speed would be determined by the kinetic energy $U$ of the material moving in the field of the crack, given by

$$U = \tfrac{1}{2}\rho \iint (\dot{u}^2 + \dot{v}^2)\, dx\, dy = \tfrac{1}{2}\rho \dot{c}^2 \iint \left[ \left( \frac{\partial u}{\partial c} \right)^2 + \left( \frac{\partial v}{\partial c} \right)^2 \right] dx\, dy \tag{6.24}$$

where $\rho$ is the density and $u$ and $v$ are the displacements at the point $x, y$. Since $\partial u/\partial c \simeq \partial v/\partial c \simeq \sigma/E$ near the crack, and since the field of the crack is concentrated mainly within a distance of the order of $c$ from the centre, the integral is of the form

$$U = \tfrac{1}{2}k_1 \rho \dot{c}^2 c^2 \sigma^2/E^2 \tag{6.25}$$

where $k_1$ is a constant. This energy has to be included with the strain and surface energies in the Griffith energy balance. For a fixed total energy the speed $\dot{c}$ of the crack is then found to be

$$\dot{c} = k_2 c_0 \left( 1 - \frac{c_e}{c} \right)^{1/2} \tag{6.26}$$

where $k_2 = (4\pi/k_1)^{1/2}$, $c_0$ is the speed of longitudinal elastic waves in a rod, i.e. $(E/\rho)^{1/2}$, and $c_e$ is the equilibrium crack length, i.e. $\gamma E/\pi \sigma^2$.

Using a specific model, Roberts and Wells (1957) calculated $k_2 = 0.38$, which agrees reasonably with the observed upper limiting speeds of long cracks in brittle bodies (Gilman, 1959), which average $0.31\ c_0$. A similar theoretical value has been derived by Stroh (1957) using the argument that a long moving crack is mechanically similar to a Rayleigh surface wave.

We expect maximum speeds in practice to be smaller than these theoretical values, for the following reason. When a running crack becomes very long relative to the Griffith length for the applied stress acting on it, its growth releases elastic energy in amounts much greater than are needed for the work of fracture $2\gamma$ in the brittle mode. It thus becomes possible for the fracture to change to some other mode, less economical in work of fracture, for example by roughening of the fracture faces or forking of the crack.

## 6.8 INTRINSIC BRITTLENESS AND DUCTILITY

We now examine the assumption that the mode of fracture shown in fig. 6.6(b) is preferred to any other for a tensile crack. In the network of highly strained atomic bonds at the tip of the crack there are many possible modes of failure, according to the various different ways in which the atoms might separate by pulling apart or by sliding over one another. Each such mode has its own characteristic strength $\sigma_m$ in the sense that it could become active if the appropriate component of the stress field of the crack were to reach this value. In the simplest cases, one mode fails clearly before the others, due to the combination of a low $\sigma_m$ and a favourable orientation of the stress field, and by its failure it then prevents the stress from rising to the higher levels needed to activate the others. The atomic mechanism of fracture is thus determined by the microscopically weakest mode of failure at the crack tip, for the orientation of crack and applied stress in question. This microscopically weakest mode need not be, and in fact often is not, the same as the macroscopically weakest mode of fracture because the macroscopic strength of a cracked body depends on $b_m$ as well as $\sigma_m$ (cf. (6.17)). The material may, for example, choose a microscopically ductile mode of failure in which its Griffith strength is large, but it is then trapped in this microscopically weak mode and cannot switch to a more brittle microscopic mode in which its Griffith strength is small. Much of the useful strength and toughness of engineering metals is due to the weakness of their microscopically ductile modes of failure.

Along the $x$-axis, ahead of the crack tip, there is biaxial tension and plane strain, i.e. $\sigma_{xx} = \sigma_{yy} = \sigma_{zz}/2\nu \leqslant \sigma_m$ so that, roughly, the ratio of maximum tensile stress to maximum shear stress in this region is $2/(1 - 2\nu)$, i.e. typically between 3 and 6. If the ratio of ideal tensile fracture strength to ideal shear strength is much greater than this ratio of stresses we expect the material to be intrinsically ductile, i.e. to fail by plastic shear rather than brittle cracking even at the end of the sharpest slit; and similarly, in the reverse case we expect it to be intrinsically brittle.

Copper, silver, and gold, because of their low elastic rigidities on close-packed slip planes and their ability to form partial dislocations on those planes, have particularly low ideal shear strengths, only about 0.03 of their ideal tensile fracture strengths, and they are of course consistently ductile. At the other extreme, diamond, with its structure of directed covalent bonds, has a shear strength nearly equal to its tensile strength and is intrinsically brittle (Kelly, 1966). Many crystalline solids have intermediate values and are less easily classified, in either theory or practice, as brittle or ductile. In general, the simple argument developed above is too rough to be reliable. For example, on this argument we would expect NaCl, with a tensile/shear stress ratio of 3 and a strength ratio of order unity, to be extremely brittle. However, when the effect of the large hydrostatic stress upon the cohesive properties is taken into account, the ideal tensile strength is increased by about 40 per cent above its value for uniaxial tension, and the ideal shear strength is decreased by about 50 per cent. The brittleness of NaCl is thus marginal, in agreement with the fact that single crystals of this material cleave with a work of fracture equal to the surface energy at low temperatures and with a higher value at higher temperatures (Kelly, Tyson and Cottrell, 1967).

When the tensile/shear stress ratio is only slightly smaller than the strength ratio the material may still in practice appear quite brittle, for two reasons. Firstly, because small amounts of plastic deformation near the tip of a slit may change the stress state in a direction favouring tensile fracture. Experiment has shown that, for consistent ductility, the material must be able to deform plastically on several independent glide systems, so satisfying von Mises' criterion for general deformation, and must also be able to undergo cross-slip (Groves and Kelly, 1963; Stoloff and Davies, 1964). The second reason for brittleness is that, once the crack does begin to run, there is less time for thermal fluctuations to create glide dislocations at its tip, so that in this sense the ideal shear strength is increased. An additional, complicating, factor is the effect of foreign atoms, from the environment, adsorbed on the crack faces. These may reduce the effective surface energy and give an apparently low ideal tensile strength. This is a form of stress–corrosion cracking in which even a normally ductile material may fail by a slowly running brittle crack in certain environments.

The choice of cleavage plane in a brittle crystal is closely related to the above considerations. If all crystal planes are given equal opportunities to crack, we expect that the cleavage plane will be that with the smallest $\sigma_m$. We recall that

$$\sigma_m \simeq (E\gamma/b)^{1/2} \qquad (6.27)$$

where $E$ is Young's modulus along the normal to the cleavage planes of surface energy $\gamma$ and spacing $b$. It appears in fact that the material constants that determine the observed cleavage planes of crystals are Young's modulus, surface energy, and interplanar spacing (Gilman, 1959).

Intrinsically brittle solids are often ductile in practice. Given a sufficiently high hydrostatic pressure in the applied stress system, to prevent cracks from opening, even substances such as sapphire can be deformed plastically at room temperature, as Bridgman (1952) has shown. This effect is important in geology and metalworking. Many processes such as rolling, swaging, and hydrostatic extrusion, make use of hydrostatic pressure to prevent the opening-up of cracks.

A normally brittle crystal can be ductile even in simple tension if it contains no cracks and if there is no way in which cracks can be formed in it. We shall discuss the nucleation of cracks later; in general it requires sharp, localised, stress concentrations such as the corners of surface steps or groups of glide dislocations piled up in slip or twin bands at obstacles. As an example of such ductility, we recall that NaCl crystals can be made ductile by washing them in water to dissolve surface steps (the Joffé effect) and then drying them carefully (Stokes, Johnston and Li, 1960). A normally brittle crystal may also behave in a ductile manner when it contains freely mobile glide dislocations (Mott, 1956). Temperature is important in this case because glide dislocations in intrinsically brittle crystals generally have large Peierls–Nabarro stresses and so are freely mobile only at high temperatures. Mott recognised that mobile dislocations can be moved by the stress field of a propagating crack and so dissipate some of its kinetic energy. As the crack then begins to slow down, more time is allowed for further movement of dislocations, so that the crack is slowed down still more and brought to rest in a shower of dislocations. At low temperatures this may not happen and the crack can then propagate rapidly in a brittle manner. In this way Mott explained the brittle–ductile temperature transition which is such a striking feature of many crystalline materials. A good demonstration of the effect of temperature on the fracture of steel was provided by Robertson's (1953) experiment in which a brittle crack, started at low temperature, was propagated at high speed up a temperature gradient in a large stressed plate until it reached the 'crack-arrest' temperature where it turned into a ductile failure and stopped. Etch-pit studies on ionic crystals have shown that dislocations can be generated prolifically at places where cracks have slowed down or stopped (Gilman and Johnston, 1956; Tetelman and Robertson, 1963).

## 6.9 MAGNITUDE OF THE GRIFFITH STRESS

We now consider numerical values of the Griffith stress. For a general guide we take $8\gamma = \mu b$, where $\gamma$ is the surface energy and $b$ is the atomic spacing. With $A = \mu/2\pi(1 - \nu)$, (6.17) then becomes

$$\frac{\sigma}{\mu} \simeq \left(\frac{b}{4c}\right)^{1/2} \tag{6.28}$$

so that, under ideally brittle conditions, microscopic cracks (e.g. $c \simeq 10^{-4}$ cm) can spread at stresses of order $10^{-2}$ $\mu$, which explains the familiar weakness of brittle objects. This macroscopic weakness does not of course imply any corresponding microscopic weakness but is due to the smallness of the fracture displacement $b_m$, which causes $\gamma$ to be small. Typically, $\sigma_m \simeq \mu/4$ and $b_m \simeq b$ for ideally brittle fracture.

To test his theory, Griffith broke thin pieces of glass which contained cracks of various known lengths and showed that the tensile strength varied as $c^{-1/2}$, as expected. The coefficient of proportionality also agreed well with the theoretical value, based on the known elastic constant of glass and the measured surface energy of molten glass. The strength of common glassware is about $10^{9}$ dyne cm$^{-2}$, which implies the presence of cracks about $10^{-4}$ cm deep. Such cracks can be displayed on the surface of glass by the technique of decoration with sodium vapour, originated by Andrade and Tsien (1937) and subsequently developed (Argon, 1959; Gordon, Marsh and Parratt, 1958). They may be caused by abrasion, chemical attack and surface devitrification. The spectacular increase in the macroscopic strength of glass, to about $3 \times 10^{10}$ dyne cm$^{-2}$ ($\simeq E/20$), which occurs when they. are removed by smoothing the surface, is now familiar (Thomas, 1960; Morley, 1963).

Recent work has suggested that the numerical agreement obtained by Griffith may have been fortuitous. When glass is really dry, its Griffith stress rises to a value which corresponds to an effective surface energy some fifty times higher than the true value (Shand, 1961), a result which Marsh (1964) has shown to be consistent with

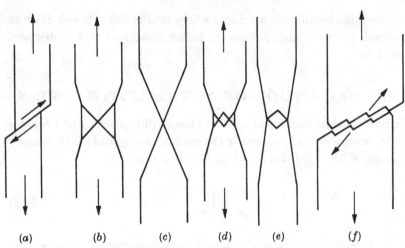

(a)    (b)   (c)   (d)   (e)    (f)

Fig. 6.7. Simple types of ductile failure

several other pieces of evidence in indicating that plastic deformation may occur at the tip of the crack. The brittle fracture of glass may thus be due to a microscopically ductile mode of cracking. In ordinary atmospheres the higher strength is not observed, presumably because water vapour enters the crack and chemically attacks the highly strained molecular network at the tip.

There is a great deal of evidence that the fracture strengths of most materials, and even the modes of fracture, are sensitive to environmental effects. Silica fibres are about four times stronger if baked and tested in a vacuum instead of air. Orowan (1944) argued that adsorption of molecules from the atmosphere would lower the effective surface energy and hence the fracture strength. In support of this idea he showed that the apparent surface energy of mica, as determined by Obreimow's fracture method, was about 4500 erg $cm^{-2}$ in a vacuum and 375 $erg cm^{-2}$ in air. Such a change would produce about a three-fold change of Griffith stress.

The idea that materials can be weakened by adsorbed molecules has been applied to many problems of fracture, and also exploited practically in the drilling of hard quartz rock by adding a trace of aluminium chloride, a surface-active substance, to the drilling water (Lichtmann, Rehbinder and Karpenko, 1958; Nabarro, 1955).

## 6.10   DUCTILE FAILURE

A ductile body can be separated into parts by concentrated plastic deformation. The basic mode of failure here is simple shear along a slip line, as in fig. 6.7(a). The energy consumed as plastic work in the active slip line is usually enormously greater than the increases of surface and elastic energy associated with the offsets at the ends of the slip line. The Griffith criterion in this case reduces simply to the requirement that the applied stress be equal to the yield stress for this mode of plastic deformation. The plastic deformation in the active slip line is very great, so that work hardening plays a decisive part in determining the onset of shear failure. The deformation cannot concentrate in one slip line unless the coefficient of work hardening has become nearly zero or even negative (work softening). Contrary to what is often stated, ductile failure is not usually due to an 'exhaustion of ductility', but to an exhaustion of work hardening capacity. Simple shear failure occurs mainly in ductile metals and alloys that have been intensely hardened, for example by cold-working or alloy precipitation. The plastic instability is often accompanied and intensified by adiabatic softening (Zener, 1948a), in which the heat released by the plastic work, in the active slip line, locally softens the material. The slip line usually follows a crystallographic glide plane in a single crystal, and a surface of plastic instability in a polycrystal (Hill, 1950; Thomas, 1961).

Thin wires, strips, or sheets of very hard ductile materials sometimes fail suddenly by shear when used as highly stressed tensile members in engineering structures. Even when the failure is completely ductile the macroscopic effect resembles that of a brittle fracture. The elongation produced by the shear is, at most, of the same order as the thickness $t$ of the material. When the length $l$ of the member is such that $t/l$ is smaller than the elastic strain, the elastic elongation of the member is more than sufficient to accommodate the failure elongation. A capacity for work hardening, usually measured in engineering tests by the ratio of ultimate tensile stress to initial yield stress, is an important practical property of ductile materials since it can prevent the plastic strain from concentrating until after a uniform plastic elongation, orders of magnitude greater than the elastic elongation, has occurred.

A simple variant of shear failure (Onat and Prager, 1954) is shown in fig. 6.7(b) and (c). Slip occurs equally and symmetrically along two intersecting orthogonal slip lines, so that the material necks down, ideally to a knife-edge failure. We notice that the shear on each slip line passes material through the other slip line and that this material is sheared by the other slip line as it does so. The shear strain in each slip line is thus unity. A finite volume of material is uniformly strained in the necked region even though, at any instant during the process, only those thin layers of material in the slip lines at that instant are actually deforming. Because this strain is less concentrated, necking failure can occur at a higher coefficient of work hardening than simple shear failure.

One of the most familiar types of ductile failure is the axi-symmetric necking of a round tensile bar, which sets in at the point on the tensile true stress–strain $(\sigma, \epsilon)$ curve where the coefficient of work hardening, $d\sigma/d\epsilon$, becomes equal to the tensile stress $\sigma$, for plastic deformation at constant volume. Ideally, the bar necks down to a sharp point.

In practice there is usually less than total reduction of area at the narrowest part of the neck when a bar breaks. The reason for this in ductile materials is that internal holes form and expand by plastic deformation. Fig. 6.7(d) and (e) show a simple example, in plane strain deformation, in which a diamond-shaped hole, connected to active slip lines at its side corners, is enlarged by simultaneous slip along these slip lines (Orowan, 1949). A double knife-edge failure then results; or, in axi-symmetric conditions, a double cone failure. The usual convention in engineering tests is to measure the reduction of area from the size of the external neck and there is then less than total reduction of area, even though the material itself may be completely ductile. When failure occurs by the growth and coalescence of a large number of small holes, as in fig. 6.6(a), a macroscopically flat but microscopically dimpled surface of separation, i.e. a fibrous fracture, is formed. The central region of a ductile tensile failure is often of this type (Cottrell, 1959; Crussard, 1959).

Failure by the plastic growth and linking of closely spaced holes is a common cause of low elongations in ductile metals, from which various practical difficulties ensue, particularly in mechanical working operations. At their narrowest points the ligaments between closely spaced holes may be so thin that the work hardening capability of the material may be insufficient to prevent the immediate necking of these ligaments as soon as plastic deformation begins. This is the basis of the poor mechanical properties of incompletely sintered metals prepared from powder compacts; and also of metals which are porous, for example because of the evolution of gases during casting. Foreign particles such as globules of slag are also a source of plastic holes in metals. When the metal begins to deform plastically, such a particle may break, due to the load thrown upon it, or the interface between the particle and the metal may tear open. In either case a hole is then created which can grow by plastic deformation.

Holes often grow during simple shear. They convert the failure from the ideal 'sliding-off' type shown in diagram (*a*) of fig. 6.7 to the more typical 'void sheet' type of diagram (*f*), in which separation

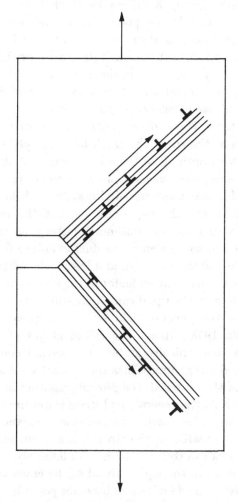

Fig. 6.8. Growth of a notch under plane strain conditions by a non-cumulative process

occurs after a much smaller overall shear (Rogers, 1960). The conical rim of the familiar tensile cup-and-cone failure appears to form in this way.

An important aspect of failure by the plastic growth of holes is that this process enables notches and cracks to spread, under certain conditions, by plastic deformation at applied stresses less than the general yield stress. Without such holes, the plastic deformation is non-dilatational and the failure mode of fig. 6.5 is therefore impossible, under plane strain conditions. (Thin sheets can deform under plane stress conditions in which local thinning may occur, in the region ahead of the crack, giving a failure mode that can be formally represented by fig. 6.5.) Under plane strain conditions, then, and without holes to provide dilatation, the dislocations which represent the process of plastic failure at the crack tip must move in a way that avoids dilatation, for example along shear lines as in fig. 6.8. Such a mode of failure is fundamentally different from that of fig. 6.5 since the contribution of each dislocation to the growth of the crack is now non-cumulative. These dislocations are associated with the sliding-off process at the tip of the crack but their glide along their slip lines contributes nothing to the advancement of the crack. To continue this advance, more dislocations have to be created continually at the tip and injected into the slip lines. These dislocations push their predecessors along the slip lines, so that the plastic zones spread outwards across the material, much faster than the crack tip spreads. This has two consequences for ductile failure from a notch. First, although the failure can begin at a low applied stress, because of the elastic stress concentration factor of the notch which produces local plastic yielding at the tip, it cannot continue unless the stress is increased towards the general yield stress, to spread the plastic zones (McClintock, 1958). Hence the initial stages of this plastic crack growth are mechanically stable. The second consequence is that, as the applied stress approaches the general yield stress of the notched section of the material, the growing plastic zones reach the far side of the body. At the general yield stress the entire load-bearing section, from the tip of the crack to the far side, is traversed by yield zones. Dislocations created at the tip can then run out at the far surface and the crack can continue to grow indefinitely.

When holes expand in the region ahead of the crack tip, a dilatational and cumulative mode of failure becomes possible, such as that of fig. 6.6(a). The equations of the Griffith theory can be applied to such fractures (Orowan, 1949; Irwin, 1948), with $2\gamma$ now representing the plastic work of deformation and separation of the bridges between the holes. Although the $\sigma_m$ factor in the expression for $2\gamma$ (equation (6.19)) is now greatly reduced (e.g. to the order of $10^{-2} \mu$) compared with its value for ideally brittle fracture, this is usually amply compensated for by the very much higher value of $b_m$ which results from the fact that the elementary process of failure, the plastic shear or necking of a bridge, is of microscopic dimensions, e.g. $10^{-2}$cm, instead of atomic dimensions. Much deeper cracks are then required to weaken the material. For example, with these values of $\sigma_m$ and $b_m$ we have $R_0 \simeq 1$ cm from (6.15); a crack of this order of depth would weaken the material by only about 30 per cent (equation (6.23)).

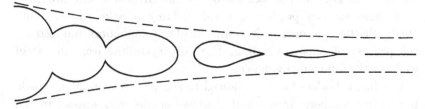

Fig. 6.9. Cumulative fracture by discontinuous brittle cracking followed by plastic failure

Whether a plane strain ductile fracture will spread by the cumulative mode of fig. 6.9 given that holes are present, or by a non-cumulative mode such as that of fig. 6.8, depends on the microscopic strengths of the two modes. In favour of the cumulative mode is the fact that the cross-sectional areas of the bridges between the holes are small. Against it is the fact that the localised deformation of these bridges occurs under plastic constraint which in extreme cases can raise the effective tensile yield stress of the bridging material by a factor of about three (Orowan, 1946). An example of plastic constraint at bridges seems to occur in the macroscopically brittle fracture of large steel plates at temperatures in the range of $0\,^\circ$C (Tipper, 1957). Brittle cleavages form separately in the grains ahead of the main crack front and then join up by plastic deformation of the material in the ligaments between them. We expect that, because it occurs under plastic constraint, this linking up will only be preferred to a non-cumulative mode of failure if at least two-thirds of the cross-section in the preparatory zone is first broken by cleavage (Cottrell, 1963). In fact, there is evidence to suggest that the crack-arrest temperature in ships' plates is that at which two-thirds of the fracture is cleavage (Hodgson and Boyd, 1958).

## 6.11  PROMOTION OF BRITTLE FRACTURE BY PLASTIC DEFORMATION

So far we have developed the view, as in the classical theories of Ludwik (1926) and Davidenkow (1936), that plastic deformation and brittle cracking are independent alternative modes of failure of a solid. This is broadly true, but is too simple to cover all aspects of brittle fracture. There is much evidence that plastic deformation in fact often helps brittle cracks form and grow. We have already mentioned the evidence for glass. In single crystals of MgO it has been shown that small cleavage cracks, $10^{-4}$ to $10^{-3}$ cm deep, in the surface, can grow at stresses far below the Griffith stress, provided that these stresses produce general yielding (Clarke and Sambell, 1960). Brittle fractures in iron and steel are commonly induced by the process of plastic yielding. For polycrystalline iron and steel we have the following evidence:

1. When a Lüders band is allowed to run part way along a tensile bar at low temperatures, small cleavage cracks may appear in the plastically yielded part but not in the unyielded part (Hahn, 1959).

2. When the yield stress is increased, by rapid straining, ageing, or irradiation, the brittle fracture stress often increases by the same amount (Churchman et al., 1957; Owen, 1959).

3. The brittle fracture stress often coincides with the yield stress over a range of temperature and grain size (Low, 1956).

Brittle fractures in steel have however been observed at stresses well below the yield stress (Allen, 1959). In some cases these are not cleavages but fractures along grain boundaries weakened by oxidation, corrosion, or segregated impurities (McLean, 1962). As regards low-stress cleavage fractures it is important to distinguish between the yield stress for general yielding and that for local yielding in a region of concentrated stress. Experiments with sharply notched bars at low temperatures have shown that brittle cleavage is preceded by deformation twinning, and deformation twinning is preceded by slip, in the grains at the root of the notch. The overall deformation produced by this localised slip and twinning is too small to detect by the conventional methods used for determining the general yield stress (Knott and Cottrell, 1963; Griffiths and Cottrell, 1965).

How then does plastic deformation cause brittle fracture: by making cleavage cracks or by helping them to grow? Orowan (1959) suggested a process for the second of these in which gliding dislocations running across the material would, as they passed near the tip of a small crack, bring their own stress fields to bear on that crack and help the applied stress make it grow. The slow growth of short cleavage cracks at the yield stress in MgO crystals is fairly certainly due to a process of this type. It has been calculated that about 300 dislocations with favourably oriented Burgers vectors are required within about $3 \times 10^{-4}$ cm from the tip of the crack (Smith, 1966).

Following the original suggestion by Zener (1948b) that slip bands could create cracks at obstacles where they produce stress concentra-

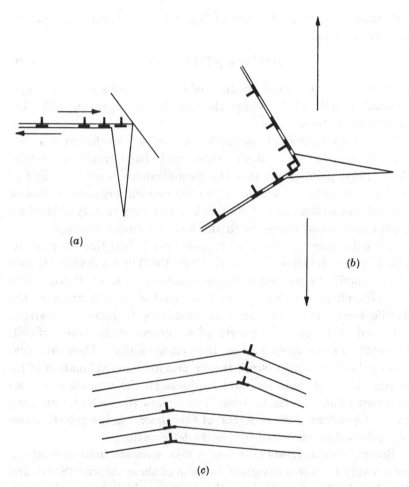

Fig. 6.10. Cleavages produced by crystal dislocations

tions, there have been various suggestions for processes in which glide dislocations may create cracks, as in fig. 6.10, and much evidence has been obtained for the occurrence of such processes in practice. (For a summary, see Cottrell, 1963.)

Particularly striking is the observation of Honda (1961) that single crystals of silicon-iron, when pulled along the [110] axis at low temperatures, formed cleavage cracks *parallel* to the tensile axis after plastic deformation had produced slip on the following systems:

(a) $(101)[\bar{1}\bar{1}1]$ and $(10\bar{1})[111]$ which intersect along $[010]$,
(b) $(011)[\bar{1}\bar{1}1]$ and $(0\bar{1}1)[111]$ which intersect along $[100]$.

According to the mechanism of fig. 6.10(b), acting through the dislocation reaction

$$\tfrac{1}{2}a[\bar{1}\bar{1}1] + \tfrac{1}{2}a[111] = a[001], \tag{6.29}$$

these intersections could produce edge dislocations of $a[001]$-type, favourably oriented to wedge cleavage cracks open on $(001)$, i.e. parallel to the tensile axis.

The theory that cleavage cracks in steel are nucleated by glide dislocations piled up in slip or twin bands has developed actively. Mott (1956) pointed out that the discontinuous plastic yielding of steel at low temperatures must produce sudden large avalanches of dislocations, so that the stress could build up very rapidly at piled-up groups and hence might reach the level to create cleavage cracks before subsequent plastic deformation nearby had time to relax it. This idea was developed by Stroh (1955, 1957) into a detailed theory of the ductile–brittle temperature transition in steel. Petch (1958) and Cottrell (1958) then pointed out that several features of the ductile–brittle transition, such as sensitivity to hydrostatic stress, indicated that the early growth of a microcrack is more difficult, i.e. requires more applied stress, than its nucleation. Their modification of the Zener–Mott–Stroh theory provided an explanation of the strong effects of grain size and hardness on the transition. It also provided a link with the Ludwik–Davidenkow theory by interpreting the 'independent fracture stress' of this theory as the growth stress of a microcrack nucleated by plastic deformation.

Recent work suggests that cracks may nucleate from dislocations more easily than was envisaged in some of these theories (Smith and Barnby, 1967). Nevertheless, the factors which make the early growth of a microcrack more difficult than its nucleation have not yet been completely identified. In a process such as that of fig. 6.10(b) and equation (6.29), in which the screw components of the two sets of dislocations attract one another, it is possible under certain circumstances for these dislocation attractions themselves to provide the driving force to nucleate the microcrack; and this microcrack then needs some applied stress for its growth. However, if the dislocation sources in the intersecting slip or twin bands continue to create more dislocations, replacing those that have entered the crack, the forces from these additional dislocations should enable the crack to continue growing. In this case growth becomes harder than nucleation only if the microcrack meets some barrier as it advances, which increases the work of fracture in the Griffith equation. The most obvious such barrier is a grain boundary and there are many published examples of small cracks in steel which appear to have stopped at grain boundaries. Furthermore, when the brittle fracture strengths of polycrystalline steel are fitted to (6.17), with $c$ set equal to the grain size, a work of fracture more than ten times larger than the true surface energy is usually obtained (Low, 1954). It has been interpreted as the plastic work required to rupture the boundary and allow the crack to break out into the next grain.

The observation of short stable cracks formed by plastic deformation in single crystals, as in the work of Honda (1961) and Hull (1960), shows that grain boundary obstacles are not the only cause of stopped cracks. We are forced back either to the process mentioned at the beginning of the previous paragraph, or to some inhomogeneity within a crystal that can act as a site for easy nucleation of a crack when stressed by the dislocations in a slip or twin band (McMahon and Cohen, 1965; Smith and Barnby, 1967). An obvious possibility is a foreign particle, such as a carbide, sulphide, or a silicate inclusion, which is brittle and has a low work of fracture, or adheres weakly to the matrix.

## 6.12 THERMAL FRACTURES

We turn now to consider fractures in which the atomic mechanism is provided by thermal or chemical action – diffusion, evaporation, dissolution – leaving the applied stress to satisfy the thermodynamic condition only. In such cases even blunt notches can grow and, if the stress contributes by means of elastic work, as in (6.20), they can grow even under compressive stresses.

Consider a narrow elliptical slit of major axis $2c$ normal to the principal stress $\sigma$, inside a body. Let this slit change shape slowly by the migration of atoms from its ends to its sides. Since an atomistic process of migration can operate slowly in a quasi-reversible way, with negligible dissipation of energy, the energy balance governing the change in shape of the slit is precisely the same as that in Griffith's theory, i.e. work is done by the applied forces acting through the elastic deformation of the slit and is converted into surface energy (equation (6.20)). Hence (6.17) gives the stress above which the ellipse will elongate and below which it will contract. At the Griffith stress, which may be tensile or compressive, the chemical potential of an atom is the same at the ends and sides of the slit and there is no net flow in either direction.

The situation changes radically when the atomic migrations provide other means for the applied stress to do work, e.g. through Nabarro–Herring creep. Consider short slits on a grain boundary normal to a small tensile stress $\sigma$. By the above argument, the slits spheroidise. However, remaining as spheroids, they can still grow by a process of lattice or grain boundary diffusion in which atoms migrate from their surfaces on to the boundary and so cause the adjoining grains to grow in the direction of the tensile stress. The removal of a thin surface layer, of volume $\delta V$, of material from a spherical hole of radius $r$ increases the surface energy of the hole by $2\gamma\delta V/r$, and its deposition in the grain boundary causes the applied stress to do the work $\sigma\delta V$. The hole can thus grow at a stress greater than (Baluffi and Seigle, 1957)

$$\sigma = 2\gamma/r. \tag{6.30}$$

This stress can be very small, e.g. $2 \times 10^7$ dyne cm$^{-2}$ (300 psi) for a hole of radius $10^{-4}$ cm with a specific surface energy of 1000 erg cm$^{-2}$. This is because the process is extremely effective in extracting useful work from the applied stress. For example, in an incremental growth of a hole the Nabarro–Herring work is of the order of $E/\sigma$ times larger than the elastic work. Of course, plastic processes of crack growth, such as that of fig. 6.8, also extract large amounts of work from the applied stress, but they dissipate most of this in the slip lines; the diffusion process, on the other hand, is quasi-reversible when operating slowly under a small concentration gradient of vacancies.

Because of its effectiveness and efficiency, this diffusion process is expected to be dominant in the limit of prolonged low stresses and high temperatures. Since the observations of Greenwood, Miller and Suiter (1954), it has become clear that during slow creep at high temperatures polycrystalline metals often fracture by 'cavitation', in which spheroidal holes appear at spacings of a few microns along grain boundaries, slowly grow, and coalesce. In at least some cases, as for example when magnesium alloys cavitate under stresses of about 100 psi at 450 °C, there can be little doubt that the diffusion process is responsible. Calculations have shown that it can operate sufficiently fast to account for observed rates of cavitation (Hull and Rimmer, 1959). At high temperatures and low stresses the holes form fairly randomly on most grain boundaries inclined at large angles to the main tensile axis and there is no obvious notch sensitivity. In agreement with this we note that, if $b_m \simeq r \simeq 10^{-4}$ cm and $\sigma_m \simeq 2\gamma/r \simeq \mu b/4r$, the preparatory zone for a cavitation crack of the form shown in fig. 6.6($a$) would be about 2 cm long (cf. (6.15)).

At lower temperatures and higher stresses grain boundary cavities usually become irregular in shape, as if produced by plastic tearing. At higher stresses, of course, additional processes such as plastic deformation, which are less efficient but faster than the diffusion processes, can come into operation. Grain boundary sliding plays an important part in many, if not all, intercrystalline creep fractures. It appears to create the nuclei for cavitation fractures, perhaps by a process similar to that of fig. 6.7($f$), the cavities forming at irregularities in the boundaries. In at least some examples, these irregularities are provided by foreign particles (Cottrell, 1961; Boettner and Robertson, 1961; Gifkins, 1965).

At temperatures somewhat below those for cavitation, grain boundary sliding produces 'triple-point cracking' in which cracks are nucleated at the junctions of differently oriented grain boundaries and then slowly grow as well-defined cracks along boundaries transverse to the tensile axis (Zener, 1948$a$; Chang and Grant, 1956$a$, $b$; Eborall, 1956). Fig. 6.10($b$) shows a typical form of such a fracture if we interpret the two inclined slip lines as sliding grain boundaries and the crack as spreading along a third boundary normal to the tensile stress.

There are obvious similarities between triple-point cracking and the cleavage process of fig. 6.10($b$). The conclusion that the crack nucleates easily applies equally to both (Smith and Barnby, 1967). A short grain boundary crack cannot grow rapidly, of course, because its growth is controlled by the rate at which the triple-point wedge is driven into it by the sliding grain boundaries. As McLean (1956–7) has pointed out, however, triple-point cracks do grow to sizes at which Griffith's criterion is satisfied. It is interesting that, even at this size, the cracks continue to grow slowly. Presumably a crack tip, slowly spreading along a grain boundary, is blunted by diffusion or plastic deformation and must depend on a thermal process to provide a mechanism for parting its atoms.

## 6.13 CHEMICAL EFFECTS IN FRACTURE

Chemical effects play an important part in many fractures. At one extreme there are chemical heterogeneities of microstructure – e.g. foreign particles of brittle or weakly adherent second phases in the material, brittle films and segregated layers along grain boundaries, soft 'denuded' zones along grain boundaries and slip lines in precipitation-hardened alloys – which can nucleate cracks either directly, or indirectly by providing sliding interfaces which set up stress concentrations. At the other extreme, external agents from the environment can interact with the stressed material and affect its mechanical behaviour in various ways. We have already referred to the adsorption of foreign atoms which can reduce the effective surface energies of crack faces and so reduce the fracture strength.

The embrittlement of steel by hydrogen is very important practically. At room temperature atomic hydrogen has a solubility in iron of about $5 \times 10^{-4}$ ppm and a diffusion coefficient of about $10^{-5} cm^2 sec^{-1}$. The metal can be readily supersaturated with hydrogen by electrolysis, acid pickling, or exposure to furnace atmospheres at high temperatures, and often then becomes brittle, particularly during slow straining at or near room temperature. The fact that strain rate has an effect is evidently a result of a diffusion process in the events leading to embrittlement. In some cases there is little doubt that hydrogen migrates to holes in the metal and forms molecular gas there, the pressure of which forces open small cracks. There are also good reasons for believing that hydrogen assists brittle fracture by adsorption on crack faces. The ensuing reduction of surface energy has been estimated from the Langmuir isotherm; it amounts, for example, to 900 erg cm at 4 $cm^3$ $H_2$ per 100 $cm^3$ Fe (Petch and Stables, 1952; Petch, 1956).

Many metals are embrittled by contact with liquid metals such as mercury, sodium, bismuth, lead and gallium, which produce intercrystalline fractures. A simple consideration of surface energies explains much of the effect (McLean, 1961). For brittle fracture along grain boundaries the effective surface energy to be used in the Griffith equation is

$$\gamma_{eff} = \gamma - \tfrac{1}{2}\gamma_{GB}$$

where the true surface energy $\gamma$ is usually about three times the grain boundary energy $\gamma_{GB}$. An adsorbed substance at the grain boundaries generally reduces both $\gamma$ and $\gamma_{GB}$, with the usual result that $\gamma_{eff}$ is reduced and the metal weakened against intercrystalline fracture. In extreme cases, for example aluminium in liquid gallium, the metal falls apart along its grain boundaries even without applied stress. In such examples we have $\gamma = \gamma_{SL}$ and $2\gamma_{SL} < \gamma_{GB}$, i.e. the liquid metal flows along the grain boundaries and parts the grains, replacing one grain boundary ($\gamma_{GB}$) by two solid–liquid interfaces ($2\gamma_{SL}$). In less extreme examples, an applied stress is needed but the metal may nevertheless be very weak and brittle. An important feature here is that $\gamma_{eff}$ can be reduced from its normal value – which may be large enough to protect the metal from brittle fracture, by forcing plastic deformation to occur at tips of cracks – to a low value at which plastic flow is not produced, in which case the Griffith equation with this reduced value of $\gamma_{eff}$ then determines the fracture strength.

Recent work has shown that plastic deformation is sometimes necessary to start the embrittlement of a metal by a liquid metal (Kamdar and Westwood, 1965). It is possible that the grain boundary cracks are started by an attack of the liquid metal on dislocations piled up in slip bands, against grain boundaries at the surface.

Stress–corrosion cracking, i.e. the weak and brittle fracture of normally strong and ductile alloys in certain chemical environments, is both scientifically challenging and technologically important. Intergranular stress–corrosion cracking is usually considered to be caused by a selective and concentrated electrochemical dissolution of grain boundaries which have become anodic by the segregation of alloy elements and impurities. In this case the role of the stress may be simply to pull the grains away from one another, as they become disconnected by intergranular corrosion, and so open up internal boundaries to the attack.

A wider role of stress – which may be a self-stress, as in the classical season-cracking of cold-worked brass in ammoniacal atmospheres – is however indicated by the occurrence of transgranular stress–corrosion cracking in certain alloys. Of course, in a ductile material, once a crack has been formed there is the likelihood that plastic stretching of the tip, with an accumulation of dislocations, will there expose a highly anodic metal surface suitable for concentrated chemical attack (Hoar and Hines, 1956a, b). There remains, however, the problem of forming the crack in the first place.

Electron microscopy has recently made important contributions to this subject. Experiments on copper alloys showed that transgranular cracking seems to occur only in alloys of low stacking fault energy. Subsequent work has suggested that the essential requirement is the concentration of slip dislocations in well-defined planes which give coarse slip steps on the surface. These slip steps appear to break open a protective surface film and this enables corrosion pits then to form on them. By electrochemical dissolution these pits develop into deep tunnels along the slip planes, and the tunnels then expand sideways by plastic deformation and necking of the material between, until they coalesce to form a continuous crack (Swann and Embury, 1965).

## 6.14 METAL FATIGUE

Because of its importance and intractability, the problem of metal fatigue has received great attention, particularly from experimentalists who have felt, not unreasonably, that the processes involved are too complex to be unravelled except by direct observation. Such observations have shown that a fatigue crack in a ductile polycrystalline metal under push–pull stressing typically forms and grows in two distinct stages. It starts at the outer surface, in the outcrop of a prominent slip band, and spreads round the surface and into the metal, along this slip band and also along conjoint bands in neighbouring grains, to form a shear slit some two or three grains deep, inclined at about $45°$ to the principal stress. This stage 1 starts early in the fatigue test and continues for nearly all of the total fatigue life in tests which last some $10^6$ cycles. The transition to stage 2 is usually fairly abrupt. The crack changes direction sharply and grows, at a rate much faster than before, normal to the principal stress.

There are thus five things to explain: (a) the formation of prominent slip bands, (b) the initiation of surface cracks in such bands, (c) the mechanism of stage 1 growth, (d) the transition to stage 2, and (e) the mechanism of stage 2 growth. Only for the last of these do we have anything like a widely agreed picture at the present time. In stage 2 the surface of the crack becomes rippled with fine lines parallel to the line of the crack front. By means of fatigue tests in which the amplitude of the load cycle was varied systematically and then correlated with corresponding variations in the ripple markings, it was proved that one ripple forms during each cycle and indicates the incremental growth of the crack in that cycle (Forsyth and Ryder, 1961). A careful examination of the crack tip at various phases of the cycle, in high-strain fatigue tests, then showed that the crack grows by plastic stretching of the tip during the tensile phase, so creating a new crack surface, followed by the folding together of the upper and lower parts of this new surface during the compressive phase, so deepening the crack and sharpening its tip again in readiness for the next tensile phase. Provided the newly formed surfaces do not weld when they are folded together – and this does not seem to happen in practice, perhaps because of adsorption effects – this is a geometrically irreversible process and the crack continues to grow, incrementally during each cycle (Laird and Smith, 1962). As regards the magnitude of each increment, we might expect that a plastic displacement at the tip, as given by equations (6.7) to (6.9), would determine this. It has been shown, however, that the observed increments are much greater, at least in high-strain fatigue (Bilby, Cottrell, Smith and Swinden, 1964). Presumably they are determined through a balance between the strain-concentrating action of the notch and the work hardening of the material.

The earlier parts of the fatigue process are much less well understood. An explanation of the formation of prominent slip bands may be developed in terms of the Bauschinger effect, work softening, and the development of dislocation sub-grain structures (Ham, 1966). In some precipitation-hardened alloys the softening of slip bands through the destruction of the precipitation-hardened state by alternating plastic deformation is well established.

Two experimental observations which have stimulated ideas about the formation of the surface crack, in prominent slip bands, are that the process can occur at very low temperatures and hence is fundamentally mechanical rather than thermal in nature, and that it seems to be related to the extrusion of material out of the slip bands at the surface, which suggests that the crack starts as a geometrical irregularity, i.e. it starts by the shuffling of different layers of the slip band in and out of the surface. Mott (1958) showed how a screw dislocation, oscillating from end to end of the slip band during the stress cycle and cross-slipping from one side of the band to the other at the ends of its path, could slide the tongue of material, enclosed by its circuit, in or out of the slip band at the surface. This proposed mechanism has since been supported by experiments which have shown that the ability and opportunity to cross-slip correlate with ease of formation of fatigue cracks (Avery, Millar and Backofen, 1961). Whether the fatigue crack grows during stage 1 by a continuation of this same type of process (McEvily and Boettner, 1963), or by one similar to that in stage 2 (Laird, 1967), or by the formation and linking up of pores in the slip bands (Wood, 1965), is not yet clear. The transition from stage 1 to stage 2 of crack growth is the least investigated of all parts of the fatigue process. Presumably, when the crack reaches a certain depth its strain-concentrating ability becomes sufficient to enforce the stage 2 mode of plastic failure, despite the greater plastic constraint compared with the stage 1 mode.

## 6.15  CONCLUSION

It is clear that fracture, as a field of study, possesses those features which typify all the many scientific fields that have attracted Professor Sir Nevill Mott's interest; fields in which explanations are demanded, at the atomic level, of bulk properties of matter often of great technological importance; and in which the problems are so complex that only a simultaneous joint attack by theory and experiment has any real prospect of success.

# AUTHOR INDEX AND BIBLIOGRAPHY

Aarts, W. H., *see* Molenaar, J.
Abelès, F. (1953), *Compt. Rend.
Acad. Sci. Paris* **237**, 796       26
Adams, M. A. and Cottrell, A. H.
(1955), *Phil Mag.* **46**, 1187       211
Aerts, E., Delavignette, P., Siems,
R. and Amelinckx, S. (1962), *J.
Appl. Phys.* **33**, 3078       148
Ahearn, J. S., *see* Labusch, R.
Ahlers, M. and Haasen, P. (1962),
*Acta Met.* **10**, 977       192, 234
Alexander, H. and Haasen, P.
(1967), *Can. J. Phys.* **45**, 1209       192
— and Mader, S. (1962), *Acta Met.*
**10**, 887       192
— *see* Schäfer, S.
Allen, J. W. (1956), *J. Electron.
Control* **1**, 580       157
Allen, N. P. (1959), In *Report of
Swampscott Conference on Frac-
ture*, ed. B. L. Averbach *et al.*
(John Wiley, New York and
MIT Press, Cambs., Mass.
(joint publ.)) p. 123       270
Amelinckx, S. (1967), *Dislocations
and Mechanical Properties of
Crystals* (John Wiley, New
York) p. 3       44
—, Bontinck, W., Dekeyser, W.
and Seitz, F. (1957), *Phil. Mag.*
**2**, 355       96
— *see* Aerts, E.
— *see* Art, A.
— *see* Gevers, R.
— *see* Siems, R.
— *see* Strumane, R.
— *see* Van Landuyt, J.
Andrade, E. N. da C. and Tsien,
L. C. (1937), *Proc. Roy. Soc.
A* **159**, 346       263
Ardley, G. W. (1955), *Acta Met.* **3**,
525       181
— and Cottrell, A. H. (1953), *Proc.
Roy. Soc. A* **219**, 328       181
Argon, A. S. (1959), *Proc. Roy.
Soc. A* **250**, 472       263
— and Brydges, W. T. (1968),
*Phil Mag.* **18**, 817
210, 212, 213, 218, 223
— and East, G. (1968), *Trans.
Japan. Inst. Metals* **9**, 756       215, 217
Art, A., Gevers, R. and Amelinckx,
S. (1963), *Phys. Stat. Sol.* **3**, 697       148
*see* Gevers, R.

Ashby, M. F. (1966), *Phil. Mag.* **14**,
1157       186, 193, 194, 224, 229, 230
— (1971), In *Strengthening
Methods in Crystals*, ed. A.
Kelly and R. B. Nicholson
(Applied Science Publishers
Ltd, London) p. 137       193, 230
— and Brown, L. M. (1963a), *Phil.
Mag.* **8**, 1083       125, 148
—, — (1963b), *Phil. Mag.* **8**, 1649 125, 148
— and Johnson, L. A. (1969), *Phil.
Mag.* **20**, 1009       85
— and Smith, G. C. (1960), *Phil.
Mag.* **5**, 298       225
— *see* Ebeling, R.
Atkinson, J. D., Brown, L. M. and
Stobbs, W. M. (1973), In *Proc.
3rd Internat. Conf. on Strength
of Metals and Alloys*, Vol. I
(Inst. of Metals and Iron and
Steel Inst., London) p. 36       230
Authier, A. and Lang, A. R. (1964),
*J. Appl. Phys.* **35**, 1956       53
Avery, D. H., Millar, G. A. and
Backofen, W. A. (1961), *Acta
Met.* **9**, 892       280
Averbach, B. L., *see* Hahn, G. T.

Backofen, W. A., *see* Avery, D. H.
Bailey, A. I. (1957), In *2nd Int.
Cong. Surface Activity*, Vol. III,
406       255
Bailey, J. E. and Hirsch, P. B.
(1960), *Phil. Mag.* **5**, 485       144, 212
Baluffi, R. W. and Seigle, L. L.
(1957), *Acta Met.* **5**, 449       274
— *see* Simmons, R. O.
Bardeen, J. and Herring, C. (1952),
*Imperfections in Nearly Perfect
Crystals* (John Wiley, New
York) p. 261       96
Barenblatt, G. I. (1962), *Advances
Appl. Mech.* **7**, 55       250
Barnby, J. T., *see* Smith, E.
Barnes, R. S. and Mazey, D. J.
(1960), *Phil. Mag.* **5**, 1247       143
— *see* Mazey, D. J.
Basinski, S. J., *see* Basinski, Z. S.
Basinski, Z. S. (1959), *Phil. Mag.*
**4**, 393       211–213, 231, 232
— (1962), In *Proceedings of the
5th International Congress for
Electron Microscopy (Philadel-
phia)* (Academic Press, New
York) p. B13       144, 147

Basinski, Z. S. (1964), *Disc. Faraday Soc.* **33**, 93                          191, 197
— and Basinski, S. J. (1964), *Phil. Mag.* **9**, 51              209, 210, 212, 223
— and Weinberg, F. (1967), *Proc. Intern, Conf.* on Deformation of Crystalline Solids, Ottawa 1966 (Ottawa-NRC)*Can. J. Phys.***45**, No. 2, Parts 2 and 3, 453–1249       190
— *see* Jackson, P. J.
— *see* Nabarro, F. R. N.
Bassett, G. A., *see* Lotz, B.
Bauerle, J. E. and Koehler, J. S. (1957), *Phys. Rev.* **107**, 1493      30–33
Benbow, J. J. and Roesler, F. C. (1957), *Proc. Phys. Soc.* **70**, 201      254
Berner, R. (1960), *Z. Naturforschung* **15a**, 689       236
— *see* Diehl, J.
Bever, M. B., *see* Robinson, P. M.
Bevitt, E., *see* Irvine, W. H.
Biggs, W. D. and Broom, T. B. (1954), *Phil. Mag.* **45**, 246      181
Bilby, B. A. (1950), *Proc. Phys. Soc.* **A63**, 191      15, 37
—, Cottrell, A. H., Smith, E. and Swinden, K. H. (1964), *Proc. Roy. Soc.* **A279**, 1      279
—, Cottrell, A. H. and Swinden, K. H. (1963), *Proc. Roy. Soc.* **A272**, 304      173, 250, 252
Boček, M. (1963), *Phys. Stat. Sol.* **3**, 2169      190, 192
—, Hötzsch, G. and Simmen, B. (1964), *Phys. Stat. Sol.* **7**, 833      191
— and Kaska, V. (1964), *Phys. Stat. Sol.* **4**, 325      191
—, Lukáč, P., Smola, B. and Švábová (1964), *Phys. Stat. Sol.* **7**, 173      191
Boettner, R. C. and Robertson, W. D. (1961), *Trans. Met. Soc. AIME* **221**, 613      275
— *see* McEvily, A. J.
Bollmann, W. (1956), *Phys. Rev.* **103**, 1588      98, 126
— *see* Whelan, M. J.
Bol'Shanina, M. A., *see* Panin, V. Ye.
Bontinck, W., *see* Amelinckx, S.
Booker, G. R. and Brown, L. M. (1965), *Phil. Mag.* **12**, 1315      145
— and Tunstall, W. J. (1966), *Phil. Mag.* **13**, 71      150
Boser, O., *see* Seeger, A.
Bowen, D. K., Christian, J. W. and Taylor, G. (1967), *Can. J. Phys.* **45**, 903      191, 192, 201

Boyd, G. M., *see* Hodgson, J.
Bragg, W. L. (1948), In *Symposium on Internal Stresses* (Inst. of Metals, London) p. 221      56
Brandon, D. G. and Wald, M. (1961), *Phil. Mag.* **6**, 1036      26
Bridgman, P. W. (1952), *Studies in Large Plastic Flow and Fracture*, (McGraw-Hill, New York)      262
Brooks, H. (1955), In *Impurities and Imperfections* (Amer. Soc. Met., Metals Park, Ohio) p. 1      24
Broom, T. B., *see* Biggs, W. D.
Bross, H., *see* Seeger, A.
Brown, E. *see* Johnson, R. A.
Brown, L. M. (1964), *Phil. Mag.* **10**, 441      76, 145
— (1967), *Can. J. Phys.* **45**, 893      66
— (1973), *Acta Met.* **21**, 879      230
— *see* Ashby, M. F.
— *see* Atkinson, J. D.
— *see* Booker, G. R.
— *see* Shaw, A. M. B.
— and Stobbs, W. M. (1971a), *Phil. Mag.* **23**, 1185      230
—, — (1971b), *Phil. Mag.* **23**, 1201      230
— *see* Vidoz, A. E.
Brown, N. (1959), *Phil. Mag.* **4**, 693      181
— *see* Marcinkowski, M. J.
— *see* Green, H.
Brown, T. J., *see* Kocks, U. F.
Brydges, W. T. (1967), *Phil. Mag.* **15**, 1079      212, 213
— *see* Argon, A. S.
Bullough, R. and Newman, R. C. (1960), *Phil. Mag.* **5**, 921      226
Burgers, J. M. (1939), *Proc. K. ned. Akad. Wet.* **42**, 293      45, 57, 126
— (1940), *Proc. Phys. Soc.* **52**, 23      126
Byrne, J. G., Fine, M. E. and Kelly, A. (1961), *Phil. Mag.* **6**, 1119      184–186

Cahn, R. W. (1967), In *Local Order in Solid Solutions*, ed. J. B. Cohen (Gordon and Breach, New York) p. 178      163, 180
— *see* Piercy, G. R.
Carrington, W., Hale, K. F. and McLean, D. (1960), *Proc. Roy. Soc.* **A259**, 203      144
Castaing, R. and Henry, L. (1962), *Compt. Rend. Acad. Sci. Paris* **255**, 76      104
—, — (1964), *J. Microscopie* **3**, 133      104

Chadderton, L. T. (1965), *Radiation Damage in Crystals* (Methuen, London) 152
Chalmers, B., *see* Johnson, L.
Chang, H. and Grant, N. J. (1956a), *Trans. Met. Soc. AIME* **206**, 544 275
—, — (1956b), *Trans. Met. Soc. AIME* **206**, 1241 275
Chen, H. S., *see* Kocks, U. F.
Chevrier, J. C., *see* Mitchell, J. W.
Chou, Y. T. and Louat, N. (1962), *J. Appl. Phys.* **33**, 3312 173
Christian, J. W. (1970), *Proc. 2nd Internat. Conf. on Strength of Metals and Alloys* (Amer. Soc. Met., Metals Park, Ohio) p. 29 215
— and Masters, B. C. (1964), *Proc. Roy. Soc. A***281**, 223, 240 183
— *see* Bowen, D. K.
— *see* Sharp, J. V.
— *see* Taylor, G.
Churchman, A. T., Mogford, I. L. and Cottrell, A. H. (1957), *Phil. Mag.* **2**, 1271 270
Clarebrough, L. M. and Hargreaves, M. E. (1959), *Progr. Met. Phys.* **8**, 1 190, 192
— and Head, A. K. (1969), *Phys. Stat. Sol.* **33**, 431 76
—, Humble, P. and Loretto, M. H. (1967), *Can. J. Phys.* **45**, 1135 76
—, Segall, R. L. and Loretto, M. H. (1966), *Phil. Mag.* **13**, 1285 140
—, Segall, R. L., Loretto, M. H. and Hargreaves, M. E. (1964), *Phil. Mag.* **9**, 377 140
— *see* Loretto, M. H.
Clarke, F. J. P. and Sambell, R. A. J. (1960), *Phil. Mag.* **5**, 697 270
Cockayne, D. J. H., Jenkins, M. L. and Ray, I. L. F. (1971), *Phil. Mag.* **24**, 1383 146
—, Ray, I. L. F. and Whelan, M. J. (1969), *Phil. Mag.* **20**, 1265 117, 145
— *see* Jenkins, M. L.
— *see* Ray, I. L. F.
Cohen, J. B. and Fine, M. E. (1962), *J. Phys. Radium* **23**, 749 180
— *see* Hahn, G. T.
— *see* McMahon, C. J.
Coiley, J. H., *see* Smallman, R. E.
Conrad, H. (1963), *The Relation between the Structure and Mechanical Properties of Metals* (HMSO, London) p. 475 183

Cosslett, V. E. (1950), *Introduction to Electron Optics* (Clarendon Press, Oxford) 104
Cotterill, R. M. J. (1961), *Phil. Mag.* **6**, 1351 139, 140, 151
— (1963), *Phil. Mag.* **8**, 1937 151
— and Segall, R. L. (1963), *Phil. Mag.* **8**, 1105 138, 140
Cottrell, A. H. (1948), In *Report of a Conference on the Strength of Solids* (The Physical Society, London) p. 30 37, 86
— (1952), In *L'Etat Solide, 9me Conseil de Physique Solvay* (Stoops, Brussels) p. 487 161
— (1953), *Dislocations and Plastic Flow in Crystals* (Clarendon Press, Oxford) p. 39 67, 161
— (1954), *Relation of Properties to Microstructure* (Amer. Soc. Met. Metals Park, Ohio) p. 131 159, 180, 181
— (1956), *Metallurgical Reviews* **1**, 479 135
— (1957), *Dislocations and Mechanical Properties of Crystals* (John Wiley, New York) p. 509 89
— (1958), *Trans. Met. Soc. AIME* **212**, 192 272
— (1959), In *Report of Swampscott Conference on Fracture*, ed. B. L. Averbach *et al.* (John Wiley, New York and MIT Press, Cambs., Mass. (joint publ.)) p. 20 266
— (1961), In *Iron and Steel Inst. and Inst. Metals Joint Symposium on Structural Processes in Creep* (Iron and Steel Inst., London), p. 1 275
— (1963), *Proc. Roy. Soc. A***276**, 1 269, 271
— (1963a), *The Relation between the Structure and Mechanical Properties of Metals* (HMSO, London) p. 455 161, 163
— (1963b), *The Relation between the Structure and Mechanical Properties of Metals* (HMSO, London) p. 566 183
— (1964), *Theory of Dislocations* (Blackie, Glasgow and London) p. 62 255
— and Bilby, B. A. (1949), *Proc. Phys. Soc. A***62**, 49 41, 161
— and Stokes, R. J. (1955), *Proc. Roy. Soc. A***233**, 17 211, 212
— *see* Adams, M. A.

Cottrell, A. H. see Ardley, G. W.
— see Bilby, B. A.
— see Churchman, A. T.
— see Griffiths, J. R.
— see Kelly, A.
— see Knott, J. F.
— see Piercy, G. R.
— see Tyson, W. R.
Crawford, R. C., Ray, I. L. F. and
   Cockayne, D. J. H. (1973),
   Phil. Mag. 27, 1                         146
— see Ray, I. L. F.
Crump, J. C. and Young, F. W.
   (1968), Phil. Mag. 17, 381               209
Crussard, C. (1959), In Report of
   Swampscott Conference on Frac-
   ture, ed. B. L. Averbach et al.
   (John Wiley, New York and
   MIT Press, Cambs., Mass.
   (joint publ.) p. 524                     266
Cundy, S. L., Metherell, A. J. F.
   and Whelan, M. J. (1965), J. Sci.
   Inst. 43, 712                            104
—, Metherell, A. J. F., Whelan,
   M. J., Unwin, P. N. T. and
   Nicholson, R. B. (1968), Proc.
   Roy. Soc. A307, 267                      104

Damask, A. C. and Dienes, G. J.
   (1963), Point Defects in Metals
   (Gordon and Breach, New
   York)                                    2, 32
—, Dienes, G. J. and Vineyard,
   G. H. (1959), Phys. Rev. 113, 781        2
Darwin, C. G. (1914), Phil. Mag.
   27, 315, 675                             122
Dash, W. C. (1957), Dislocations
   and Mechanical Properties of
   Crystals (John Wiley, New
   York) p. 57                              96
Davidenkow, S. N. (1936), Dina-
   mickeskaya Ispytania Metallov,
   Moscow                                   269
Davidge, R. W. and Pratt, P. L.
   (1964), Phys. Stat. Sol. 6, 759          191
Davies, R. G. and Stoloff, N. S.
   (1965), Phil. Mag. 12, 297               187
— see Johnston, T. L.
— see Stoloff, N. S.
Dehlinger, U. (1929), Annln. Phys.
   (5) 2, 749                               43
Dekeyser, W., see Amelinckx, S.
Delavignette, P. and Vook, R. W.
   (1963), Phys. Stat. Soli. 3, 648         144
— see Aerts, E.
— see Siems, R.
De Sorbo, W. (1960), J. Phys.
   Chem. Sol. 15, 7                         32

Dew-Hughes, D. (1960), Acta Met.
   8, 816                                   184
Diehl, J. and Berner, R. (1960), Z.
   Metallkunde 51, 522                      236
— and Hinzner, F. (1964), Phys.
   Stat. Sol. 7, 121                        215
— see Seeger, A.
Dienes, G. J. and Vineyard, G. H.
   (1957), Radiation Effects in
   Solids (Interscience, New York)          152
— see Damask, A. C.
Dietze, H. D., see Leibfried, G.
Dobson, P. S., Goodhew, P. J. and
   Smallman, R. E. (1967), Phil.
   Mag. 16, 9                               145
— and Smallman, R. E. (1966),
   Phil. Mag. 14, 357                       145
Dorn, J. E., see Guyot, P.
— see Mitchell, J. B.
Drum, C. M. (1965), Phil. Mag. 11,
   313                                      115, 148
— and Whelan, M. J. (1965), Phil.
   Mag. 11, 205                             148
Dudarev, Ye. F., see Panin, V. Ye.
Duesbery, M. S. and Foxall, R. A.
   (1969), Phil. Mag. 20, 719              192
—, Foxall, R. A. and Hirsch, P. B.
   (1966), J. de Physique 27,
   C3-193                                  191, 192
— and Hirsch, P. B. (1970), In
   Proc. of Conference on Funda-
   mental Aspects of Dislocation
   Theory, ed. Simmons, de Wit
   and Bullough (Nat. Bur. of
   Standards, Washington) p.
   1115                                     225
— see Foxall, R. A.
Dupouy, G. and Perrier, F. (1962),
   J. Microscopie 1, 167                    101

East, G., see Argon, A. S.
Ebeling, R. and Ashby, M. F.
   (1966), Phil. Mag. 13, 805       193, 224
Eborall, R. (1956), In Symposium
   on Creep and Fracture of Metals
   at High Temperatures (HMSO,
   London) p. 229                           275
Edington, J. W. and Smallman,
   R. E. (1965) Phil. Mag. 11,
   1109                                     76, 145
Edmondson, B. and Williamson,
   G. K. (1964), Phil. Mag. 9, 277          147
Embury, J. D., see Swann, P. R.
Eshelby, J. D. (1949), Proc. Phys.
   Soc. Lond. A62, 307                      50
— (1953), J. Appl. Phys. 24, 176           46

Eshelby, J. D. (1956), In *Progress in Solid State Physics*, ed. Seitz and Turnbull (Academic Press, New York) **3**, 78      34, 42
— (1961), *Phil. Mag.* **6**, 935      164
— (1966), *Brit. J. Appl. Phys.* **17**, 1131      53
—, Frank, F. C. and Nabarro, F. R. N. (1951), *Phil. Mag.* **42**, 351      84, 128
Essmann, U. (1963), *Phys. Stat. Sol.* **3**, 932      191, 197, 205
— (1964), *Acta Met.* **12**, 1468      202, 245
— (1965a), *Phys. Stat. Sol.* **12**, 707      191, 197, 205, 244
— (1965b), *Phys. Stat. Sol.* **12**, 723      191, 197, 205, 244
— *see* Rühle, M.
Eyre, B. L. (1973), *J. Phys.* **F3**, 422      143

Feltner, C. E. (1966), *Phil. Mag.* **14**, 1219      147
Fine, M. E., *see* Byrne, J. G.
— *see* Cohen, J. B.
Fischbach, D. B. and Nowick, A. S. (1958), *J. Phys. Chem. Sol.* **5**, 302      89
— (1953), *Phys. Rev.* **91**, 232      180
— (1954), *Acta Met.* **2**, 9      180
—, Hart, E. W. and Pry, R. H. (1953), *Acta Met.* **1**, 336      185
Fisher, R. M. (1959), *Rev. Sci. Instr.* **30**, 925      104
— (1962), Thesis, Cambridge University      163
— *see* Marcinkowski, M. J.
Flanagan, W. F. (1967), Private communication      183
Fleischer, R. L. (1962a), *Acta Met.* **10**, 835      155
— (1962b), *J. Appl. Phys.* **33**, 3504      184
— (1963), *Acta Met.* **11**, 203      157, 158
— (1966), *Acta Met.* **14**, 1867      157
— (1967), *Acta Met.* **15**, 1513      183
— and Hibbard, W. R. (1963), *The Relation between the Structure and Mechanical Properties of Metals* (HMSO, London) p. 262      155, 165, 170
Flinn, P. A. (1958), *Acta Met.* **6**, 631      159
— (1960), *Trans. Met. Soc. AIME* **218**, 145      187
Foreman, A. J. E. (1955), *Acta Met.* **3**, 322      61
—, Hirsch, P. B. and Humphreys, F. J. (1970), In *Proceedings of*

*Conference on Fundamental Aspects of Dislocation Theory*, ed. Simmons, de Wit and Bullough (Nat. Bur. of Standards, Washington) p. 1083      227
—, Jaswon, M. A. and Wood, J. K. (1951), *Proc. Phys. Soc.* **A64**, 156      51, 253
— and Makin, M. J. (1966), *Phil. Mag.* **14**, 911      170, 171, 214, 226, 232
Forsyth, P. J. E. and Ryder, D. A. (1961), *Metallurgia* **63**, 377      279
Fourie, J. T. (1967), *Can. J. Phys.* **45**, 777      197
— and Wilsdorf, H. G. F. (1959), *Acta Met.* **7**, 339      195
— *see* Wilsdorf, H. G. F.
Foxall, R. A., Duesbury, M. S. and Hirsch, P. B. (1967), *Can. J. Phys.* **45**, 607      191, 192, 193, 196, 197, 201, 202
— *see* Mitchell, T. E.
Frank, F. C. (1950), *Carnegie Institute of Technology Symposium on the Plastic Deformation of Crystalline Solids* (Office of Naval Research, Washington) p. 150      136
— (1955), In *Report of the Conference on Defects in Crystalline Solids* (Physical Society, London) p. 159      126
— (1963), *Metal Surfaces* (Amer. Soc. Met., Metals Park, Ohio) p. 1      66
— and Stroh, A. N. (1952), *Proc. Phys. Soc.* **B65**, 811      86
— *see* Eshelby, J. D.
Frank, W. (1967a), *Z. Naturf.* **22a**, 365      155
— (1967b), *Z. Naturf.* **22a**, 377      164
— (1967c), *Phys. Stat. Sol.* **18**, 459      155
— (1967d), *Phys. Stat. Sol.* **19**, 239      183
Frenkel, J. and Kontorova, T. (1938), *Phys. Z. Sowj. Un.* **13**, 1      7, 48
—, — (1939), *Fiz. Zh.* **1**, 137      48
Friedel, J. (1952), *Phil. Mag.* **43**, 153      25
— (1955), *Phil. Mag.* **46**, 1169      235
— (1963a), *Electron Microscopy and Strength of Crystals* (Interscience, New York) p. 605      165, 170
— (1963b), *The Relation between the Structure and Mechanical Properties of Metals* (HMSO, London) p. 409      183
— (1964), *Dislocations* (Pergamon Press, Oxford)      67, 86

Friedel, J. and Saada, G. (1968), In *Rept. of Metallurgical Soc. Symposium on Workhardening*, ed. J. P. Hirth and J. Weertman (Gordon and Breach, New York) p. 1    223

Friedrich, W., Knipping, P. and Laue, M. von (1912), *K. Bayer. Akad. München Ber.* p. 303    43

— — — (1913), *Ann. d. Physik* **41**, 971    43

Fumi, F. (1955), *Phil. Mag.* **46**, 1007    25

Gallagher, P. C. J. (1964), *Disc. Faraday Soc.* **33**, 157    212

— (1966), *Phys. Stat. Sol.* **16**, 95    76

— (1967), *Phil. Mag.* **15**, 51    213

— (1970), *Met. Trans.* **1**, 2429    77

Gay, P., Hirsch, P. B. and Kelly, A. (1954), *Acta Cryst.* **7**, 41    126

Gevers, R., Art, A. and Amelinckx, S. (1963), *Phys. Stat. Sol.* **3**, 1563    148

— *see* Art, A.

— *see* Strumane, R.

— *see* Van Landuyt, J.

Gifkins, R. C. (1965), In *Report of Tewkesbury Symposium on Fracture* (University of Melbourne, distrib. by Butterworths) p. 44    275

Gilman, J. J. (1959) In *Report of Swampscott Conference on Fracture*, ed. B. L. Averbach *et al.* (John Wiley, New York and MIT Press, Cambs. Mass. (joint publ.)) p. 193    259, 262

— and Johnston, W. G. (1956), In *Dislocations and Mechanical Properties of Crystals*, ed. J. C. Fisher (John Wiley, New York) p. 116    263

— *see* Johnston, W. G.

Gleiter, H. (1968), *Acta Met.* **16**, 1213    225

— and Hornbogen, E. (1965), *Phys. Stat. Sol.* **12**, 235, 251    182

Goodhew, P. J., *see* Dobson, P. S.

— *see* Tunstall, W. J.

Gordon, J. E., Marsh, D. M. and Parratt, M. E. M. L. (1958), *Proc. Roy. Soc.* A**249**, 65    263

Goringe, M. J., *see* Swann, P. R.

Gould, D., Hazzledine, P. M., Hirsch, P. B. and Humphreys, F. J. (1973), In *Proc. 3rd Internat. Confer. on Strength of*

*Metals and Alloys* (Inst. of Metals and Iron and Steel Institute, London) p. 31    230

Grange, G., *see* Labusch, R.

Grant, N. J., *see* Chang, H.

Green, H. and Brown, N. (1953), *J. Metals* **5**, 1240    181

Greenwood, J. N., Miller, D. R. and Suiter, J. W. (1954), *Acta Met.* **2**, 250    274

Griffith, A. A. (1921), *Phil. Trans. Roy. Soc.* A**221**, 163    180, 254

Griffiths, J. R. and Cottrell, A. H. (1965), *J. Mech. Phys. Sol.* **13**, 135    270

Grilhé, J. (1967), In *Physique des Dislocations* (Presses Universitaires de France)    226

Groves, G. W. and Kelly, A. (1961), *Phil. Mag.* **6**, 1527    147

—, — Erratum (1962), *Phil. Mag.* **7**, 892    147

—, — (1963), *Phil. Mag.* **8**, 877    261

Guyot, P. and Dorn, J. E. (1967), *Can. J. Phys.* **45**, 983    183

Gyulai, Z. and Hartley, D. (1928), *Z. Phys.* **51**, 378    89

Haasen, P. (1958), *Phil. Mag.* **3**, 384    236

— (1965), In *Physical Metallurgy*, ed. R. W. Cahn (North-Holland, Amsterdam) p. 821    185

Haasen, P., *see* Ahlers, M.

— *see* Alexander, H.

— *see* Labusch, R.

— *see* Schäfer, S.

Hahn, G. T. (1962), *Acta Met.* **10**, 727    161

—, Averbach, B. L., Owen, W. S. and Cohen, M. (1959), In *Report of Swampscott Conference on Fracture*, ed. B. L. Averbach *et al.* (John Wiley, New York and MIT Press, Cambs., Mass. (joint publ.)) p. 91    270

Hale, K. F., *see* Carrington, W.

Hall, C. E. (1953), *Introduction to Electron Microscopy* (McGraw-Hill, New York)    104

— and Hirsch, P. B. (1965), *Proc. Roy. Soc.* A**286**, 158    125

Ham, R. K. (1962), *Phil. Mag.* **7**, 1177    205

Ham, R. K. (1966), *Canadian Metallurgical Quarterly*, Sept.    279

Hargreaves, M. E. *see* Clarebrough, L. M.

Harris, B. (1966), *Phys. Stat. Sol.*
**18**, 715     159
Hart, E. W. (1972), *Acta Met.* **20**,
275     185, 230
— *see* Fisher, J. C.
Hartley, D., *see* Gyulai, Z.
Hashimoto, H., Howie, A. and
Whelan, M. J. (1962), *Proc. Roy.
Soc.* A**269**, 80   123, 124, 125, 148
Häussermann, F. and Wilkens, M.
(1966), *Phys. Stat. Sol.* **18**, 609   76
Hazzledine, P. M. (1966), *J. de
Physique* **27**, C3–210    219
— (1967), *Can. J. Phys.* **45**, 765
215, 217, 221
— (1968), *Phil. Mag.* **18**, 1033   240
— (1971a), *Phil. Mag.* **23**, 225   240
— (1971b), *Scripta Met.* **5**, 847   221
— and Hirsch, P. B. (1967), *Phil.
Mag.* **15**, 121   84, 235, 237, 238
— *see* Gould, D.
— *see* Pande, C. S.
— *see* Steeds, J. W.
Head, A. K. (1967), *Phys. Stat.
Sol.* **19**, 185     66
— *see* Clarebrough, L. M.
—, Loretto, M. H. and Humble,
P. (1967), *Phys. Stat. Sol.* **20**,
521     66
Heidenreich, R. D. (1949), *J. Appl.
Phys.* **20**, 993    99, 126
Henderson, J. W. and Koehler,
J. S. (1956), *Phys. Rev.* **104**, 626   33
Henry, L., *see* Castaing, R.
Herring, C., *see* Bardeen, J.
Hesse, J. (1965), *Phys. Stat. Sol.* **9**,
209     191
Hibbard, W. R. *see* Fleischer, R. L.
Hill, R. (1950), *The Mathematical
Theory of Plasticity* (Oxford)   265
Hines, J. G., *see* Hoar, T. P.
Hinzner, F., *see* Diehl, J.
Hirsch, P. B. (1957), *J. Inst.
Metals* **86**, 7     225
— (1959a), *J. Institute of Metals*
**87**, 406     128
— (1959b), In *Internal Stresses and
Fatigue in Metals*, ed. G. M.
Rassweiler and W. L. Grube
(Elsevier, Amsterdam) p. 139
213, 231, 232
— (1962), *Phil. Mag.* **7**, 67   89
— (1963), *Relation between the
Structure and Mechanical Pro-
perties of Metals* (HMSO,
London) p. 39   191, 197, 205
— (1964), *Disc. Faraday Soc.* **38**,
111     144

—, Horne, R. W. and Whelan,
M. J. (1956), *Phil. Mag.* **1**, 677
98, 99, 104
—, Howie, A., Nicholson, R. B.,
Pashley, D. W. and Whelan,
M. J. (1965), *Electron Micro-
scopy of Thin Crystals* (Butter-
worths, London)   101, 104, 110, 119,
125, 143, 147–9, 151
—, — and Whelan, M. J. (1960),
*Phil. Trans. Roy. Soc.* A**252**,
499    113, 116–118
Hirsch, P. B. and Humphreys,
F. J. (1969), *Physics of Strength
and Plasticity*, ed. A. S. Argon,
(MIT Press, Cambridge, Mass.)
p. 189    193, 225, 226
—, — (1970a), In *Report of 2nd
Internat. Conf. on Strength of
Metals and Alloys* (Amer. Soc.
Met., Metals Park, Ohio) p. 545   197
—, — (1970b), *Proc. Roy. Soc.*
A**318**, 45    225, 227, 229
— and Kelly, A. (1965), *Phil. Mag.*
**12**, 881     178
— and Lally, J. S. (1965), *Phil.
Mag.* **12**, 595   192, 195, 202, 212, 213,
215, 216–18, 220, 245
— and Mitchell, T. E. (1967), *Can.
J. Phys.* **45**, 663   191, 197, 214, 231,
232, 235, 239–43
—, — (1968), *Rept of Metallurgical
Soc. Symposium on Workhardening*, ed. J. P. Hirth and J.
Weertman (Gordon and Breach
New York) p. 65
201, 231, 235, 242, 243
—, Silcox, J., Smallman, R. E. and
Westmacott, K. H. (1958),
*Phil. Mag.* **3**, 897    136, 137
— and Warrington, D. H. (1961),
*Phil. Mag.* **6**, 735     213
— *see* Bailey, J. E.
— *see* Duesbery, M. S.
— *see* Foreman, A. J. E.
— *see* Foxall, R. A.
— *see* Gay, P.
— *see* Gould, P.
— *see* Hall, C. R.
— *see* Hazzledine, P. M.
— *see* Humphreys, F. J.
— *see* Mitchell, T. E.
— *see* Silcox, J.
— *see* Thornton, P. R.
— *see* Tunstall, W. J.
— *see* Valdrè, U.
— *see* Whelan, M. J.

Kuhlmann-Wilsdorf, D. (1968) —cont.
J. Weertman (Gordon and Breach, New York) p. 97     215, 239
— see Wilsdorf, H.

Labusch, R. (1969), Crystal Lattice Defects 1, 1     187
— (1970), Phys. Stat. Sol. 41, 659     188
— (1972), Acta Met. 20, 917     188
—, Ahearn, J. S., Grange, G. and Haasen, P. (1975), In Metals Trans., Dorn. Memorial Symp. 1972, Rate Processes in Plastic Deformation, AIME Symposium Series (Plenum Press, New York)     188
Laird, C. (1967), In 69th Meeting of Amer. Soc. Test. Mat., (Amer. Soc. Test. Mat., Philadelphia) ACM spec. tech. publ. 415, p. 131     280
— and Smith, G. C. (1962) Phil. Mag. 7, 847     279
Lally, J. S., see Hirsch, P. B.
Lang, A. R., see Authier, A.
Laue, M. von, see Friedrich, W.
Le Clare, A. D., see Makin, S. M.
Leibfried, G. and Dietze, H. D. (1949), Z. Phys. 126, 790     50, 51
Leitz, C., see Mader, S.
Levinson, L. M. and Nabarro, F. R. N. (1967), Acta Met. 15, 785     27
Levinstein, H. J. and Robinson, W. H. (1963), The Relation between the Structure and Mechanical Properties of Metals (HMSO, London) p. 179     210
Li, C. H., see Stokes, R. J.
Lichtmann, V. I., Rehbinder, P. and Karpenko, G. V. (1958), Effect of a Surface-Active Medium on the Deformation of Metals (HMSO, London)     264
Livingston, J. D. (1959), Trans. Met. Soc. AIME 215, 566     175
— (1962), Acta Met. 10, 229     210, 212
— see Phillips, V. A.
Logie, H. J. (1957), Acta Met. 5, 106     181
Loretto, M. H. (1964), Phil. Mag. 10, 467     145, 148
— (1965a), Phil. Mag. 12, 1087     145, 148
— (1965b), Phil. Mag. 12, 125     145
—, Clarebrough, L. M. and Humble P. (1966), Phil. Mag. 13, 953     140
—, — and Segall, R. L. (1964), Phil. Mag. 10, 731     145

—, —, — (1965), Phil. Mag. 11, 459     140
— see Clarebrough, L. M.
— see Head, A. K.
Lothe, J., see Jøssang, T.
Lotz, B., Kovacs, A. J., Bassett, G. A. and Keller, A. (1966), Kolloid-z.u.z. Polymere 209, 115     95
Louat, N., see Chou, Y. T.
Love, A. E. (1952), Mathematical Theory of Elasticity (Cambridge University Press, London)     16
Low, J. R. (1954), Symposium on Relation of Properties to Microstructure (Amer. Soc. Met., Metals Park, Ohio) p. 163     273
— (1956), In Deformation and flow of Solids, ed. R. Grammel (Springer-Verlag, Berlin) p. 60     270
Low, J. R., see Stein, D. F.
Lowell, J. (1972) J. Phys. F. 2, 547     188
Ludwik, P. (1926) Zeit. Ver. dtsch. Ing. 70, 379     269
Lukáč, P., see Boček, M.

MacDonald, D. K. C. (1954), In Defects in Crystalline Solids (Physical Society, London) p. 383     28, 29
McClintock, F. A. (1958), J. Appl. Mech. 25, 582     268
McEvily, A. J. and Boettner, R. C. (1963), Acta Met. 11, 725     280
McIntyre, K. G. (1967), Phil. Mag. 15, 205     143, 147
McLean, D. (1956–7), J. Inst. Metals 84, 468     275
— (1961), In Properties of Grain Boundaries, 4th Metallurgical Colloquium (Presses Universitaires de Paris) p. 85     276
— (1962) Mechanical Properties of Metals (John Wiley, New York)     270
— see Carrington, W.
McMahon, C. J. and Cohen, M. (1965) Acta Met. 13, 591     273
Mader, S. (1957), Z. Physik 149, 73     193–196, 221, 237
— (1963), In Electron Microscopy and Strength of Crystals, ed. G. Thomas and J. Washburn, (Interscience Publishers, New York) p. 183     192, 194, 221
—, Seeger, A. and Leitz, C. (1963), J. Appl. Phys. 34, 3368     236
— — and Thieringer, H. M. (1963) J. Appl. Phys. 34, 3376     192
— see Alexander, H.
— see Seeger, A.

Makin, M. J., Whapham, A. D. and
Minter, F. J. (1961), *Phil. Mag.*
**6**, 465                                         143
— *see* Foreman, A. J. E.
— *see* Sharp, J. V.
Makin, S. M., Rowe, A. H. and Le
Clare, A. D. (1957), *Proc. Phys.*
*Soc. London B***70**, 545               32
Mann, E., Jan, R. von and Seeger,
A. (1961), *Phys. Stat. Sol.* **1**,
17                                              42
— *see* Seeger, A.
Marcinkowski, M. J., Brown, N.
and Fisher, R. M. (1961), *Acta*
*Met.* **9**, 129                               160
— and Fisher, R. M. (1963), *J.*
*Appl. Phys.* **34**, 2135                  181
— — (1965), *Trans. Met. Soc.*
*AIME* **233**, 293                         187
— and Miller, D. S. (1961), *Phil.*
*Mag.* **6**, 871                               182
Marsh, D. M. (1964), *Proc. Roy.*
*Soc. A***282**, 33                           263
— *see* Gordon, J. E.
Martin, J. W., *see* Humphreys, F. J.
Massey, H. S. W., *see* Mott, N. F.
Masters, B. C., *see* Christian, J. W.
Matsuura, K. M. and Koda, S.
(1963), *J. Phys. Soc. Japan* **18**,
suppl. 1, 50                               185
Mazey, D. J., Barnes, R. S. and
Howie, A. (1962), *Phil. Mag.* **7**,
1861                                      143, 147
— *see* Barnes, R. S.
Meissner, J. (1959), *Z. Mettall-*
*kunde* **50**, 207                          236
Metherell, A. J. F. (1971), *Advances*
*in Optical and Electron Micro-*
*scopy*, Vol. IV, (Academic Press,
New York) p. 263                          104
— *see* Cundy, S. L.
Millar, G. A., *see* Avery, D. H.
Miller, D. R., *see* Greenwood, J. N.
Miller, D. S., *see* Marcinkowski,
M. J.
Miller, P. H. and Russell, B. R.
(1952), *J. Appl. Phys.* **24**, 1248      21
Minter, F. J., *see* Makin, M. J.
Mitchell, J. B., Mitra, S. K. and
Dorn, J. E. (1963), *Amer. Soc.*
*Met. Trans. Quarterly* **56**, 236        186
Mitchell, J. W. (1957), *Dislocations*
*and Mechanical Properties of*
*Crystals* (John Wiley, New
York) p. 69                                44
—, Chevrier, J. C., Hockey, B. J.
and Monaghan, J. P. (1967),
*Can. J. Phys.* **45**, 453                  195

Mitchell, T. E. (1964), *Progr. Appl.*
*Mat. Res.* **6**, 117          190, 192, 223
—, Foxall, R. A. and Hirsch, P. B.
(1963), *Phil. Mag.* **8**, 1895           190
— and Spitzig, W. A. (1965), *Acta*
*Met.* **13**, 1169                          191
— and Thornton, P. R. (1963),
*Phil. Mag.* **8**, 1127                     185
— and Thornton, P. R. (1964),
*Phil. Mag.* **10**, 315            192, 234
— *see* Hirsch, P. B.
— *see* Thornton, P. R.
Mitra, S. K., *see* Mitchell, J. B.
Mogford, I. L. *see* Churchman,
A. T.
Molenaar, J. and Aarts, W. H.
(1950), *Nature* **166**, 690             33
Monoghan, J. P., *see* Mitchell,
J. W.
Moon, D. M. and Robinson, W. H.
(1967), *Can. J. Phys.* **45**, 1017      191
Morley, J. G. (1963), *New Scientist*
**17**, 122                                 263
Mott, N. F. (1948), *Engineering*
**165**, 16                    250, 254, 259
— (1951), *Proc. Phys. Soc. B***64**, 729   126
— (1952a), *Imperfections in Nearly*
*Perfect Crystals* (Wiley, New
York) p. 173                         165, 169
— (1952b), *Phil. Mag.* **43**, 1151
192, 235, 245
— (1955), *see* Cottrell, A. H. and
Stokes, R. J. (1955)                     212
— (1956), *J. Iron and Steel Inst.*
**183**, 233                         262, 272
— (1958), *Acta Met.* **6**, 195          279
— (1960), *Trans. Met. Soc. AIME*
**218**, 962                                234
— and Jones, H. J. (1936), *The*
*Theory of the Properties of*
*Metals and Alloys* (Oxford
University Press, London)              25
— and Massey, H. S. W. (1965),
*The Theory of Atomic Collisions*,
3rd edition (Clarendon Press,
Oxford)                              25, 110
— and Nabarro, F. R. N. (1940),
*Proc. Phys. Soc.* **52**, 86              165
—, — (1948), *Report on Strength of*
*Solids* (The Physical Society,
London) p. 1          52, 58, 165, 183
Mughrabi, H. (1968), *Phil. Mag.*
**18**, 1211                      209. 214 244
— (1971) *Phil. Mag.* **23**, 897     209, 244
Müller, E. W. (1962), In *Direct*
*Observations of Imperfections*
*in Crystals*, ed. J. B. Newkirk
and J. H. W. Wernick (Interscience, New York) p. 85                26

Nabarro, F. R. N. (1940), *Proc.*
*Roy. Soc.* A175, 519                20
— (1946), *Proc. Phys. Soc.* 58, 669     165
— (1947), *Proc. Phys. Soc. Lond.*
59, 256                          50
— (1948), *Strength of Solids* (The
Physical Society, London) p. 75     70
— (1951), *Adv. in Phys.* 1, 269      24
— (1955), *Trans. South African
Inst. Elec. Eng.* 46, 221         264
— (1967), *The Theory of Crystal
Dislocations* (Clarendon Press,
Oxford)                      84, 182
— (1972), *J. Less-Common Metals*
28, 257                         188
—, Basinski, Z. S. and Holt, D. B.
(1964), *Adv. in Phys.* 13, 193
190, 192, 216, 232, 235, 239
— *see* Eshleby, J. D.
— *see* Levinson, L. M.
— *see* Mott, N. F.
Nakada, Y. and Keh, A. S. (1966),
*Acta Met.* 14, 961               237
— *see* Keh, A. S.
Neumann, P. (1971), *Acta Met.* 19,
1233                             221
Newman, R. C., *see* Bullough, R.
Nicholson, R. B., *see* Cundy, S. L.
— *see* Hirsch, P. B.
— *see* Kelly, A.
Nihoul, J., *see* Strumane, R.
Noggle, T. S. and Koehler, J. S.
(1957), *J. Appl. Phys.* 28, 53     236
Nowick, A. S., *see* Fischback, D.B.
Nutting, J., *see* Thomas, G.


Obreimow, J. W. (1930), *Proc.*
*Roy. Soc.* A127, 290             255
Onat, E. T. and Prager, W. (1954),
*J. Appl. Phys.* 25, 491          265
Orlov, A. N., *see* Vasil'yev, L. I.
Orowan, E. (1934), *Z. Phys.* 89,
634                              43
— (1944), *Nature* 154, 341          264
— (1946), *Trans. Inst. Shipbuilders
Scotland* 89, 165                269
— (1948), In *1947 Symposium on
Internal Stresses, Discussion*
(Inst. Metals, London) p. 451      165
— (1949), *Rep. Progr. Phys.* 12,
214                          266, 268
— (1959), In *Report of Swamp-
scott Conference on Fracture*, ed.
B. L. Averbach *et al.* (John
Wiley, New York and MIT
Press, Cambs., Mass. (joint
publ.)) p. 147                   270

Owen, W. S. (1959), In *Report
of Swampscott Conference on
Fracture*, ed. B. L. Averbach
*et al.* (John Wiley, New York
and MIT Press, Cambs., Mass.
(joint publ.)) p. 141             270
— *see* Hahn, G. T.


Pande, C. S. (1968), D. Phil. Thesis,
University of Oxford             205
— (1970), *Phil. Mag.* 21, 169        240
— and Hazzledine, P. M. (1971a),
*Phil. Mag.* 24, 1039     195, 202, 203
— and Hazzledine, P. M. (1971b),
*Phil. Mag.* 24, 1393    202, 204, 242
Panin, V. Ye., Dudarev, Ye. F.,
Sidorova, T. S. and Bol'-
Shanina, M. A. (1963), *Fiz.
Metal. Metalloved.* 16, 574,
*Phys. Met. Metallogr.* 16, No. 4,
71                              159
Parratt, M. E. M. L., *see* Gordon,
J. E.
Pashley, D. W., *see* Hirsch, P. B.
Peierls, R. E. (1940), *Proc. Phys.
Soc. Lond.* 52, 34                50
Perrier, F., *see* Dupouy, G.
Petch, N. J. (1953), *J. Iron Steel
Inst.* 174, 25                   162
— (1956), *Phil. Mag.* 1, 331        276
— (1958), *Phil. Mag.* 3, 1089       272
— and Stables, P. (1952), *Nature*
169, 842                        276
Pfaff, F. (1962), *Z. Metallkunde* 53,
411, 466                        236
Pfeiffer, W. and Seeger, A. (1962),
*Phys. Stat. Sol.* 2, 668         185
Phillips, V. A. (1965), *Phil. Mag.*
11, 775                         184
— and Livingston, J. D. (1962),
*Phil. Mag.* 7, 969              148
Piercy, G. R., Cahn, R. W. and
Cottrell, A. H. (1955), *Acta
Met.* 3, 331                     185
Polanyi, M. (1934), *Z. Phys.* 89, 660   43
Prager, W. *see* Onat, E. T.
Pandtl, L. (1928), *Z. angew. Math.
Mech.* 8, 85                     43
Pratt, P. L., *see* Davidge, R. W.
Pry, R. H., *see* Fisher, J. C.


Quirk, A., *see* Irvine, W. H.


Ramsteiner, F. (1966), *Mater. Sci.
Eng.* 1, 206                     191
— (1967), *Mater. Sci. Eng.* 1, 281     191
Rapp, M., *see* Seeger, A.

Yoshioka, H. (1957), *J. Phys. Soc.*
  *Japan* **12**, 628                          125
Young, F. W. and Sherrill, F. A.
  (1967), *Can. J. Phys.* **45**, 757      205
— *see* Crump, J. C.

Zener, C. (1948*a*),  *Fracturing of*
  *Metals*  (Amer.  Soc.  Met.,
  Metals Park, Ohio) p. 1       265, 275
— (1948*b*),   *Trans.  Amer.  Soc.*
  *Met.* A**40**, 3                            271

# SUBJECT INDEX

*[A dash represents the word(s) coming before the first punctuation mark in the preceding entry, or, in some obvious cases, the first of a pair of words.]*

activation energy for self-diffusion, 32
alloys, 152–88
annealing of point defects, 32, 41; kinetics of, 32, 41
anomalous absorption, 122–5
antiphase domains, 153, 160, 181, 182, 187; tubes of, 187
attractive obstacles, 169

back stress from pile-ups, 215, 236, 237, 243
Bardeen–Herring source, 96
Bauschinger effect in alloys, 186, 230; and fatigue, 279
bend contours, 111, 112
body-centred interstitial, 2
Bollmann technique, 99, 100
brittle–ductile transition, 262, 272
brittle fracture, 254; effect of hydrostatic stress on, 262; ideal, 258; and plastic deformation, 269–73; speed of, 259
Burgers circuit, 68, 69
— vector, 68–70; determination by electron microscopy, 146, 147

cavities: and creep, 274, 275; and grain boundary sliding, 275; growth of by diffusion, 274, 275; nucleation at inclusions, 266, 267
cell structure, 134, 135
chemical effects on fracture, 276–8
chemical interaction, 159
ciné techniques, 105
classical cracks, 253
cleavage planes, 261, 262
coherent precipitates, 153, 177; effect of on flow stress of, 177–9
coincidence site lattice, 81, 82
conjugate slip system, 204
cooling stage, 104, 105
Cottrell–Stokes law, 211, 212
crack-arrest, 262
cracks: classical, 253; and dislocations, 249–52, 257; elastic stresses due to, 252, 253; Griffith, 253, 254, 258, 259; and preparatory zone, 251, 258, 275; tensile, 256; width of front, 253, 258
critical slip system, 204
cross-slip, 77, 193; and annihilation of screws, 202, 216; slip trace of, 126, 127, *see also* double cross-slip

— system, 204
crowdion, 7, 142
cumulative mode of fracture, 257, 269

density of dislocations, 127, 144; of primary and secondary dislocations, 192, 197, 201, 209
diffraction contrast, 107, 108
diffuse forces, 165–8, 173
diffusion coefficient, 14, 32
dilatation interaction, 156–9
dilatational mode of fracture, 257
dipoles, 73, 74, 197, 198, 199, 201–203; effect of on work hardening, 216, 217
dislocation sources: co-operative operation, 242, 243; Frank–Read, 53, 211; operation of in Seeger's model, 237
dislocations: addition of, 77; adsorption at, 86; anisotropic elasticity theory of, 59–64; Burgers vector of, 68–70; 'carpets' of, 197; core of, 50–2; and cracks, 249–52, 257; density of, 127, 144, 209; dipoles, 73, 74; edge, 45; effect of surface on, 46; elastic energy of, 46, 56, 61; electrical resistivity of, 151; entropy of, 67, 68; and Eshelby twist, 46; forces on, 83; helix, 96, 141; image stress field of, 46; isotropic elasticity theory of, 54–7; jogs in, 89; line tension of, 65–7; locking, 161–63; locks, 77; loops, 70–73; 'mats' of, 197; methods of observation, 44; multipoles, 73, 74; networks of, 131, 132; nodes, 129–32; origin of, 85; partial, 72–3; Peierls–Nabarro model of, 50–2; precipitation at, 86; processions of in alloys, 172–4; rearrangement, 205; screw, 45; stability of, 66, 76; Volterra, 249; width of, 253
dispersion-hardened alloys: Bauschinger effect in, 186, 230; dislocation distributions 205–208; hardening due to prismatic loops in, 227–229; interaction of screws with prismatic loops in, 225–7; Orowan loops in deformed, 207; Orowan stress, 169, 180, 184; prismatic loops in deformed, 206, 208; stage I work hardening, 193, 194, 224–230
dissociated dislocations, 75, 76, 129, 132, 149, 150
— nodes, 129–32

# 300 SUBJECT INDEX